Panzers I and II

Panzers I and II

and their Variants

from *Reichswehr* to *Wehrmacht*

Walter J. Spielberger

Schiffer Military History
Atglen, PA

Scale drawings: Hilary L. Doyle
Color illustrations: Uwe Feist
Photo sources: P. Chamberlain Collection (26), Daimler-Benz AG Archives (14), Uwe Feist Archives (18), Robert J. Icks Collection (22), Ingo Kasten Archives (2), Krauss-Maffei AG (7), K. Kassbührer (1), Landsverk (1), MAN Archives (2), Werner Oswald Archives (1), H. Scultetus (2), Walter J. Spielberger Archives (66), F. Wiener Collection (1).

    The drawings in this book were kindly contributed by Mr. Hilary Doyle, who, like the author, is a member of Bellona Publication Ltd.
    Our thanks go out to Bellona for the permission to publish these drawings. They offer by far the most complete information on German and other military vehicles.
    Four-sided views in 1:76 and 1:48 scales are available through specialty shops, or directly from Bellona Publications Ltd., Badgers Mead, Hawthorn Hill, Bracknell, Berkshire-Winkfield Row 2938, England, U.K.

Book translation by Dr. Edward Force, Central Connecticut State

Book Design by Ian Robertson.

Copyright © 2007 by Schiffer Publishing.
Library of Congress Control Number: 2006935911

All rights reserved. No part of this work may be reproduced or used in any forms or by any means – graphic, electronic or mechanical, including photocopying or information storage and retrieval systems – without written permission from the copyright holder.

Printed in China.
ISBN: 978-0-7643-2624-0

This book was originally published in German under the title
Die Panzer-Kampfwagen I und II und Ihre Abarten by Motorbuch Verlag

We are interested in hearing from authors with book ideas on related topics.

Published by Schiffer Publishing Ltd. 4880 Lower Valley Road Atglen, PA 19310 Phone: (610) 593-1777 FAX: (610) 593-2002 E-mail: Info@schifferbooks.com. Visit our web site at: www.schifferbooks.com Please write for a free catalog. This book may be purchased from the publisher. Please include $3.95 postage. Try your bookstore first.	In Europe, Schiffer books are distributed by: Bushwood Books 6 Marksbury Avenue Kew Gardens Surrey TW9 4JF, England Phone: 44 (0) 20 8392-8585 FAX: 44 (0) 20 8392-9876 E-mail: Info@bushwoodbooks.co.uk. Visit our website at: www.bushwoodbooks.co.uk Free postage in the UK. Europe: air mail at cost. Try your bookstore first.

Contents

Foreword 9

Large Tractor, Rheinmetall 12
Large Tractor, Krupp 12
Large Tractor, GT 1, Daimler-Benz 12
R/R Self-propelled Mount, Horch 13
R/R Self-propelled Mount, Duerkopp 13
WD Self-propelled Mount 3.7 cm, WD, Rheinmetall 15
WD Self-propelled Mount 7.7 cm, WD, Rheinmetall 15
Light Tractor, formerly Small Tractor—Kampfwagen,
    Krupp 20
Light Tractor—Supply Vehicle, Krupp 20
Light Tractor, formerly Small Tractor—Tank,
    Rheinmetall 21
Light Tractor—3.7 cm Antitank Self-propelled
    Mount. Rheinmetall 20
50 HP Tractor for 3.7 cm Linke-Hoffmann-Busch 25
Heavy Self-propelled Mount (Weapon Carrier), Krupp 26
Small Tractor (Tank Destroyer), Krupp 27
Landsverk 30 (German designation RR 160), Landsverk 28
Landsverk 10, Landsverk 28
Battalion Leader's Vehicle, BW, Rheinmetall 28
7.5 cm Self-propelled Mount, Rheinmetall 28
LaS "Agricultural Tractor"-Krupp Tractor, LKA, Krupp 28
LaS "Agricultural Tractor"-Krupp Tractor, LKB, Krupp 28
New Vehicle with Rheinmetall Turret, NbFz,
    Rheinmetall 30
New Vehicle with Krupp Turret, NbFz, Rheinmetall 30
New Vehicle (Design), Krupp 30
3.7 cm Self-propelled Mount L/45, Rheinmetall 33

Panzerkampfwagen I and Variations

5-ton *Panzerkampfwagen* (Design), MAN 35
5-ton *Panzerkampfwagen* Prototype, Krupp 35
5-ton *Panzerkampfwagen* (Design), Henschel 35
5-ton *Panzerkampfwagen* (Design), Daimler-Benz ... 35

5-ton *Panzerkampfwagen* (Design),
  Rheinmetall-Borsig 35
Panzerkampfwagen I (MG) (Type A), I A LaS Krupp,
  various 35
Panzerkampfwagen I (MG) (Type B), I B LaS
  May, various 43
Panzerkampfwagen I with 20 mm
  Cannon (Spanish rebuilding) 51
Panzerkampfwagen I Type A with Diesel Motor
  M 601, Krupp 40
Panzerkampfwagen I Type A (Tp), I A LsA
  Krupp, various 56
Panzerkampfwagen I Type B with Diesel motor
  M 601, Krupp 43
Panzerkampfwagen I Type B (Tp), I B LaS
  May, various 56
Panzerkampfwagen I Type C or
  Panzerkampfwagen I n.A., Krauss-Maffei 57
Panzerkampfwagen I Type F, or
  Panzerkampfwagen I n.A. reinforced, Krauss-Maffei 57
Panzerkampfwagen I Type A Driving School
  Vehicle, I A LaS Krupp, various 59
Panzerkampfwagen I Type B Driving School
  Vehicle, I B LaS May, various 59
Panzerkampfwagen I (A) Ammunition Tractor, I A
  LaS Krupp, Daimler-Benz 59
Panzerkampfwagen I Type B Repair Vehicle I,
  I B LaS May, various 59

Panzerkampfwagen I Type B Engineer Vehicle I,
  I B LaS May, various 59
4.7 cm Pak (t) on *Panzerkampfwagen*
  I minus turret, I B LaS May, Alkett 62
15 cm sIG 33 on *Panzerkampfwagen* I type
  B, or Geschützwagen 1, I B LaS May, Alkett 64
Ladungsleger I, first version, I B LaS May, Talbot 67
Ladungsleger I, second version, I B LaS May, Talbot 67
Panzerkampfwagen I Type A Flamethrower
  (troop-rebuilt), I A LaS Krupp, various 67
Kleiner *Panzerbefehlswagen* Type A, 1 kl A,
  Daimler-Benz 68
Kleiner *Panzerbefehlswagen* Type B, 2 kl B,
  Daimler-Benz 70
Kleiner *Panzerbefehlswagen* Type C, 3 kl B,
  Daimler-Benz 70
Panzerkampfwagen I Command Bridgelayer, various ....
Panzerkampfwagen VK 301 (Design), Weserhütte 70
Panzerkampfwagen VK 501 (Design), Buessing-NAG 70

Panzerkampfwagen II and Variations

10-ton *Panzerkampfwagen* Prototype, LKA 2, Krupp 71
10-ton *Panzerkampfwagen* Prototype, Henschel 71
10-ton *Panzerkampfwagen* Prototype, LaS 100, MAN 71
Panzerkampfwagen II (2 cm) Type a1,
  1/LaS 100, various 71
Panzerkampfwagen II (2 cm) Type a2,
  1/LaS 100, various 74
Panzerkampfwagen II (2 cm) Type a3, 1/LaS 100, various 74

Panzerkampfwagen II (2 cm) Type b, 2/LaS 100,
    various 75
Panzerkampfwagen II (2 cm) Type c, 3/LaS 100,
    various 76
Panzerkampfwagen II (2 cm) Type A, 4.LaS 100,
    various 77
Panzerkampfwagen II (2 cm) Type B, 5/LaS 100,
    various 77
Panzerkampfwagen II (2 cm) Type C, 6/LaS 100,
    various 77
Panzerkampfwagen II (2 cm) Type F, 7/LaS 100,
    various 89
Panzerkampfwagen II (2 cm) all types (Tp),
    LaS 100, various 101
Panzerkampfwagen II (2 cm) Type D, 8/LaS 138,
    Daimler-Benz 101
Panzerkampfwagen II (2 cm) Type E, 8/LaS 138,
    Daimler-Benz 101
Panzerkampfwagen II (2 cm) all types, amphibian,
    LaS 100, various 103
Panzerkampfwagen II Type G1, G2 and G4, MAN 106
Panzerkampfwagen II Type J, MAN 106
Panzerkampfwagen II Type H—Development
    Chassis, MAN 106
Panzerkampfwagen II Type M—Development
    Chassis, MAN 106
Panzerkampfwagen II new type for Artillery
    Command and Observation, MAN-Porsche 107
Panzerkampfwagen II new type for Battle
    Reconnaissance, MAN/Porsche 107

Panzerkampfwagen II new type for Reconnaissance,
    MAN/Porsche 107
Panzerkampfwagen II new type, reinforced, MAN 108
Panzerkampfwagen II new type, reinforced,
    for Battle Reconnaissance (small body),
    MAN/Porsche/Skoda 108
Panzerkampfwagen II with 4.7 cm Pak (t) -
    suggestion, Daimler-Benz 108
Panzerkampfwagen II new type, MAN 108
Panzerkampfwagen II Type L "Luchs," also
    Panzerspähwagen II, MAN 108
Panzerkampfwagen II Type L (5 cm KwK L/60)
    "Luchs 5 cm," MAN 110
Battle Reconnaissance Vehicle "Leopard," MIAG 111
Battle Reconnaissance Vehicle (Multipurpose Tank),
    Daimler-Benz 115
Multipurpose Tank on Panzer II Basis, FAMO 116
Armored Device 13, unknown 116
Panzerkampfwagen II (F) Type A, 8/LaS 138,
    MAN/Wegmann 117
Panzerkampfwagen II (F) Type B, 8/LaS 138,
    MAN/Wegmann 117
Armored Self-propelled Mount I for 7.62 cm
    Pak 36 (r) "Marder II," 8/LaS 138, ALKETT 117
Armored Self-propelled Mount I for 7.62 cm
    FK 296 (r) "Marder II," 8/LaS 138, ALKETT 117
Armored Self-propelled Mount II for 5 cm
    Pak 38 (prototype), LaS 100, ALKETT 117
Armored Self-propelled Mount II for 7.5 cm
    Pak 40/2 "Marder II," LaS 100, ALKETT 117
5 cm Gun on *Panzerkampfwagen* II Special
    Chassis 901 (Pz Sfl 1c) MAN 121

Leichter *Panzerjäger* (Pz. Sfl 5 cm), MAN/Porsche ....
Panzerkampfwagen II Chassis for 8.8 cm Pak 41
    (design), unknown 124
Panzerkampfwagen II chassis for 7.5 cm KwK
    L/70 (design), unknown 124
Leichte Feldhaubitze 18/2 on Panzer II (Sf) chassis
    "Wespe," LaS, FAMO 124
Munitions *Selbstfahrlafette* on Panzer II chassis,
    LaS 100, FAMO 124
Geschützwagen II for 15 cm sIG 33, LaS 100, unknown 125
Geschützwagen II widened for 15 cm sIG 33
    (6-wheel running gear), unknown 128
Panzerselfstfahrlafette for sIG 33, MAN/Porsche 108
Panzerbefehlswagen II, LaS 100, various 128

Minenraumpanzer II (*Hammerschlaggerät*),
    LaS 100, Wegmann 128
Panzerkampfwagen II (Bridgelayer), LaS 100, Magirus 128
Panzerkampfwagen II Engineer Vehicle II,
    LaS 100, various 129
Panzerkampfwagen II Driving School Vehicle,
    LaS 100, various ....
Panzerkampfwagen II n.A. reinforced Recovery
    Tank, MAN 129
Panzerkampfwagen
II Fire Control/Observation Tank, LaS 100, various 128
Panzerkampfwagen II Assault Gun dummy,
    LaS 100, various 129

Technical Data 140
Bibliography 159
Abbreviations 160

Foreword

The experience gathered until the end of World War I about the use of armored vehicles certainly confirmed the effectiveness of this new means of warfare, but did not specify its use in future conflicts.

While infantry-oriented planners propagandized the heavily armored penetration tank, the proponents of the fast, light cavalry tank stood on the other side of the argument. On the basis of the vacuum brought on by the Treaty of Versailles, Germany could closely observe the developments in other countries and draw its own conclusions from them. Political limitations, technical shortcomings, and lacking means allowed only general research at that time anyway. In spite of that, out of this situation there came trains of thought that consciously strove to cut these modern war machines loose from all traditional associations. Unnoticed by most people, it was people like Fuller, de Gaulle, and Guderian who created the basis for the later armored divisions and armies. The way to them was beset with almost insuperable hindrances.

Our study tries for the first time to bring together the technical-tactical developments of those years, and thus allow a look, unfortunately not always complete, into the formative years of the German armored weapon. By utilizing all available sources, the transition from the "black days" of the *Reichswehr* to the period of German rearmament, which produced the training devices for the new armored weapon, are covered. The fact that some of the resulting vehicles, sometimes created as variations, were still in action when the war ended in 1945 shows the superiority of the original design. Also noteworthy is the fact that in the course of this development, toward the end of the twenties, the basic concept of the later so successful Panzer III and IV tanks was established.

Despite almost thirty years of research on the subject of military vehicles, this book would not have been possible without the cooperation of my friends Col. R. J. Icks and Dr. F. Wiener.

I must also express my thanks to Messrs. P. Chamberlain, H. Doyle, U. Feist, D. Hunnicutt, H. Scultetus, and A. Sohns for years of collaboration.

Comments and criticism from readers would contribute to making later editions even more complete.

Walter J. Spielberger

Armored Tracked Vehicles Developments of the *Reichswehr*, 1925 to 1934

Right after the end of World War I, Joseph Vollmer, who was responsible for the design of the first German armored vehicles, developed the tracked tractor for civilian use. The fascination of this "WD. Raupenschlepper" (German Power Plow Company, Berlin W) lay in the hands of the Hannoversche Maschinenbau AG (Hanomag). Similar vehicles were also built under license by Podeus in Wismar, and the Dinos-Werke in Berlin. The tractors were produced in two sizes, and in fact with 20 (later 25) HP (4 x 90 x 150) and 50 HP (4 x 130 x 155) motors. For running on petroleum, these powerplants were fitted with a Grätzin-Schwer oil carburetor. The tracks got their power from steering gears to the drive wheels at the back. Through the differential in between, the tracks were independent of each other in their movement. The tractor frame was carried by rollers with spring bearings, which ran inside on the rails of the tracks. By braking the left or right differential shaft, the turning of the appropriate track was slowed or brought to a stop, thereby effecting the steering of the tractor. Track drive allowed driving on soft surfaces, since the ground pressure was only about 0.5 kg/sq.cm/ Small ditches could be crossed. In 1924 a new series of tracked tractors was introduced by Hanomag.

After the armistice of 1918, Paragraph 171 of the Treaty of Versailles had declared that Germany must not own any armored vehicles, nor could any research on their development be done. The manufacture of armored vehicles of any kind was expressly forbidden.

Armored vehicles under construction had to be destroyed on command from the International Control Commission for Germany. A reconstruction (K-Wagen) was also banned. The plans developed by Vollmer for the light "LK II" tank were saved, though, taken to Sweden, and evaluated and improved there. Thus there arose, under contract from the Swedish Army, the "m. 21" light tanks that were introduced as military equipment there as of 1921. With four-man crews, these vehicles weighed 9.5 tons. They were equipped with Scania-Vabis 60 HP four-cylinder motors, and carried a machine gun as their armament. With these vehicles the later *Generaloberst* Heinz Guderian, on the occasion of a four-week command with the "Strijdsvagn Bataillon" of the Goeta Guard, made his first acquaintance with tanks.

At this time the development of new German armored vehicles was already underway, under the strictest secrecy.

Usable information on the building of armored vehicles was not available at that time. References to World War experiences were insufficient, since the vehicles of that time scarcely allowed thorough research. The Army Weapons Office, in the vacuum of 1925, worked out requirements that corresponded to concepts of the time. A gross weight of 20 tons was established. Top speed of up to 40 km/h, minimum speed of 3 km/h, overall length up to six meters, and width to 2.6 meters. The vehicle's height should not exceed 2.35 meters. Along with a wading ability of 800 mm, complete floating ability should be available. Mastery of grades up to 30 degrees, and climbing ability of up to one meter. The specific ground pressure should not amount to more than 0.5 kg/sq.cm. Imperviousness to gas was wanted. The primary weapon should have all-round firing ability.

One of the few "A 7 V" tanks remaining in Germany after the war's end is seen during the unrest of 1919. The body of this vehicle was obviously changed in order to house the machine guns better. This vehicle too had to be scrapped.

The "LK II" vehicle was supposed to go into production toward the end of the war. These vehicles were no longer in service. All the plans were taken over by Sweden.

The "Stridsvagn m.21" used by the Swedish Army was the production model of the German "LK II."

For reasons of secrecy, disguised designations for forbidden developments were introduced in 1926. Armored tracked vehicles were given the disguise name of "Tractor." To indicate the size of the vehicles, the projects were given the names "Grosstraktor" (up to 23 tons) and "Leichttraktor" (10 to 12 tons).

Orders for "Grosstraktoren" were given to the firms of Rheinmetall (Director Remberg), Krupp/Essen (Dir. Müller), and Daimler-Benz (Dr. F. Porsche and Friedrich). These firms were advised to produce only individual parts in their factories, and send the vehicle components to the Rheinmetall works in Unterlüss. Each firm had received a contract for two vehicles.

The planned development of the "Leichttraktoren" was slowed for the time being.

The first motorization program established in 1926 foresaw, among others, self-propelled mounts for the 3.7 cm Tak and 7.5 cm leFK. This development was turned over to the Krupp firm with the provision that these vehicles should also be usable as tractors for mine launchers and infantry guns. At this time, mainly for financial reasons, the greatest value was placed on acclimatization (subvention). The industry was asked to develop military vehicles in such a way that they could also, with the slightest effort, be used by the economy. This was to guarantee an adequate number of vehicles usable by the military in case of need.

The development of the "Grosstraktor" vehicles, begun in 1925, created prototypes that were produced by three firms. They were alike in every way. This picture shows the Rheinmetall vehicle.

During the summer and autumn of 1926, negotiations took place in Moscow about the establishment of a war vehicle school in Kama, near Kazan. This led to a contract on December 9, 1926. But because the Russians made no tanks available, and German tanks could not yet be delivered, it took another two years before the school in Kama could begin its training. Then it made practical training of tank specialists possible at a time when this activity was impossible in Germany.

The Horch firm also took part in self-propelled mount development, and in 1926 they began work on a wheeled/tracked vehicle that was also supposed to be usable as a load carrier for various purposes. Besides the light type, various types of heavy self-propelled mounts were called for. The Dürkopp firm in Bielefeld worked on the design for a wheeled/tracked vehicle that was meant to carry the 7.7 or 8.8 cm Flak gun. With their tracks removed, the Dürkopp vehicles could also be used as three-axled trucks. A Krupp design of a vehicle with three driven axles was also made.

The "Grosstraktor" prototype made by Krupp. The openings in the hull sides allowed dirt to be removed from the jack rollers. The small turret in back is easy to see.

The Daimler-Benz firm also built two of these prototypes. This vehicle stood at the barracks of Panzer Regiment 5 in Wünstorf for a long time.

The rear view of the vehicle shows the arrangement of the main and secondary turrets, with the engine compartment stuck in between them.

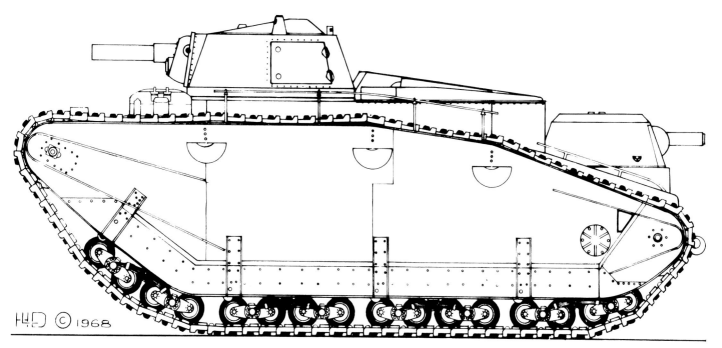

1/5 scale model of the Rheinmetall "Grosstraktor."

In 1927 the strictly secret assembly of all the "Grosstraktoren" being developed took place at the Rheinmetall works in Unterlüss.

The Rheinmetall firm was also involved in self-propelled mount development in 1927. Under contract from the *Reichswehr* Ministry, a civilian WD-Schlepper was fitted with a gun as a self-propelled mount. This stock 25 HP tractor was fitted with a 3.7 cm L/45 gun, which had a muzzle velocity of 760 meters per second. It had a traverse of 30 degrees and an elevation of -5 to +30 degrees. Then came another development with the heavy version of the WD Schlepper. Rheinmetall fitted the 50 HP WD Schlepper with a 7.7 cm gun and a coupled machine gun. The big gun's traverse was 360 degrees, its elevation from -7 to +15. The vehicle and armament were partially protected by armor.

For antitank defense, the use of the 3.7 cm *Tankabwehrkanone* L/45 was chiefly planned at that time; it could be either horse-drawn, pulled by a tractor, or mounted on a self-propelled mount for troop use. In 1927 it was determined that in the realm of Inspection 6 (K) equipment, the promotion of tests with "Tak's on Sfl" was urgent. As for production, the supplying of the planned 20 platoons with a total of 119 self-propelled Tak mounts was considered at that time.

In 1928 the first "Grosstraktoren" were completed. They were built mainly by Rheinmetall and Krupp, with the Daimler-Benz "Grosstraktor" following in 1929. Rheinmetall had developed a tracked vehicle with Cletrac tracks, which had a fighting weight of 19 tons. The installed BMW aircraft engine gave the vehicle a top speed of 40 km/h. A six-man crew was protected by 13 mm armor.

In these types, non-heat treated plates were used exclusively. The caliber of the primary weapon was 7.5 cm, and its muzzle velocity was 450 m/sec. While the traversing field was 360 degrees, the elevation field was from -2 to +60. The machine-gun elevation ran from -15 to +80 degrees. One of these vehicles, along with the two Krupp tanks, was sent to "Kama" as training equipment. The second Rheinmetall vehicle was lost in a test of its wading ability on October 30, 1929. The vehicles were used for training at Putlos in later years.

In the meantime, the assembly of the two Daimler-Benz vehicles was finished at Unterlüss.

The frame is shown with drive wheels installed.

These pictures show the assembly of the first prototype at Unterlüss in mid-January 1929. The chassis frame, made of soft steel, weighed 4193 kg.

Besides the drive wheels, the road wheels and their suspension can be seen.

The front view shows the vehicle with its protected headlight. The bow machine gun was added later.

The engine compartment cover and sidewalls are seen, looking from front to back.

The powerplant was a former aircraft engine (Type "182 206") weighing 604 kg without wheel box and starter motor.

The Central Design Bureau in Untertürkheim, under the leadership of Prof. Porsche, had created a vehicle that still much resembled the British tanks of World War I. Only one turret with 360-degree traverse was planned. The 15-ton vehicles were made of soft steel, had a self-carrying hull, and were amphibious. A former aircraft engine (Type F 182 206, built in 1918) was used, now called Type "D IV" and producing 300 HP at 1450 rpm. This six-cylinder gasoline engine with standing individual cylinders had a displacement of 31.2 liters and weighed 604 kg. A DKW two-cylinder two-stroke motor of 10 HP was seen as its starter. The self-supporting frame weighed 4193 kg. It used rear drive. The shifting of the planetary gearbox (6 forward, 2 reverse gears) and the steering worked hydraulically. The hydraulic brakes could also be used as a clutch. On each side were four road-wheel aggregates, suspended by leaf springs (wheel size (300) 90-203, weight of a wheel truck 179 kg) and hydraulic support cylinders. The tracks (Type MK 6/380/160) had a width of 380 mm, and were changed several times. The primary armament had a caliber of 7.5 cm, an overall length of 1500 mm, and was mounted in a central turret along with a heavy machine gun.

The gearbox of the Daimler-Benz vehicle. The coulisse shift lever is seen at the right. An external-band brake is attached to the driveshaft.

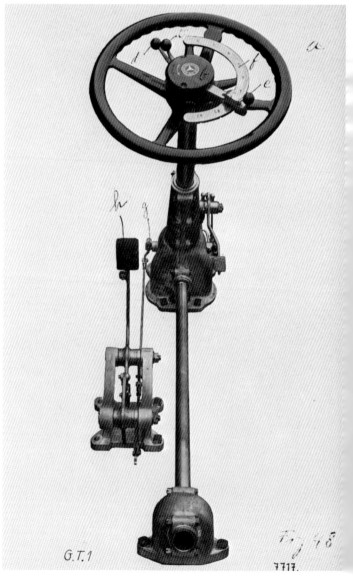

The steering of the "Grosstraktor" was done with a steering wheel.

This screw propeller was used to propel the vehicle in the water.

The tracks of the "Grosstraktor," of Type "MK 6/380/160." The rollers are easy to see. Rubber buffers are inserted on the outside. The weight per meter was 61.3 kg.

A wheel truck with leaf spring, seen before installation.

A second turret at the rear held another machine gun. The installation of a third heavy machine gun at the bow was done later, during assembly. A six-man crew was planned. The whole conception was not very good. The commander sat next to the driver in the bow of the vehicle, could not see backward at all, and was badly placed otherwise because of the protruding track ends and the low-lying seat. A later decision by Daimler Benz, of March 18, 1928, called for equipping it with radio sets while it was being built. The vehicles of the Krupp firm were laid out similarly.

Discussion at the Truppenamt on March 14, 1928, on the subject of the military vehicle program resulted in the following situation: Six test models of the "Grosstraktor" (15 tons) took place in the summer of 1928. Testing was planned for 1929 and 1930, and production was to begin in 1931. At first, equipping one company (17 vehicles) was planned, with more according to available means. The price for one company was set at about 2.5 million Marks. The delivery of the first "Kleintraktor" test models was planned for October 1929. They were to be tested from 1930 on. As of 1931 their production was planned, according to available means. At first the equipment for one company (17 vehicles) was to be made. The price per vehicle was about 50,000 Marks, so the total cost would be about a million Marks. If one "Grosstraktor" and one "Kleintraktor" company were created every year from 1931 on, the special costs would have to be decreased.

A conference of the *Wehrmacht* on 6/25/1928 concerned the armament program for motorized defensive weapons. As a basis for the ongoing work on the armament program, the need for antitank guns was determined:

A.	Tak (horse-drawn or towed by tractors)	354 units
B.	Tak (mounted on tractors)	138 units
C.	Tak guns for light tractors	34 units
D.	Tak guns for armored vehicles	40 units

In all, therefore, 566 Tak guns (3.7 cm) were required. Because of supplying 129 Tak guns more than had been scheduled with necessary ammunition, other requirements for the armament program followed. First the means had to be made clear.

These had to be calculated anew for the additional Tak guns, for ammunition for the 129 Tak, and for pivoting and installation work to convert 138 tractors to self-propelled mounts. The 7.5 cm gun program was meanwhile continuing, and 34 of these guns were planned as "Grosstraktor" weapons, while six test pieces were created for antitank defense. Heavy machine guns were produced: 138 for Tak on tractors, for light "tractors," and three apiece for heavy "tractors," for a total of 102 guns. In addition, six heavy machine guns were ordered for antitank defense.

On July 19, 1928, the Army command, in Memo IV No. 560/28 geh. Kdos. *Wehrmacht* In 6 (K), stated the final requirements for the building of the "Small Tractor." WaPrüf had been asked to issue orders to the Daimler, Krupp, and Rheinmetall firms on the basis of these plans for the design and building of two vehicles apiece. The design drawings of the three firms were to be presented for approval as soon as possible. If the Daimler firm could not accept the contract, it was advised that a third chassis be ordered from each of the other firms. One firm should concern itself with an armored supply vehicle, while the other should develop a 3.7 cm self-propelled antitank gun mount. The general requirements stated that the "Small Tractor" was a light combat vehicle. Its designation was changed to "Leichttraktor" (light tractor).

Within the motorization of the Army, the vehicle was to fulfill several tasks simultaneously. It was planned as a load carrier (self-propelled mount for 3.7 cm Tak, ammunition, and provision supply vehicle for use at the front lines, with appropriately armored bodies). It was also to be a towing tractor for light and, if possible, also medium loads. The possibility of making the chassis useful in the civilian economy was to be striven for.

For the "Leichttraktor," a number of requests were made. At first a semi-automatic 3.7 cm cannon (dimensions of a fully automatic type were to be considered) was to be joined by a heavy machine gun in the turret, with a 260-degree traverse. As great a positive and negative elevation as possible was to be attained. The smallest supplies of ammunition were set at 150 rounds for the cannon and 3000 for the MG. If more were carried, cannon ammunition had priority. Seats for four crewmen (gunner and commander in the turret, driver and radioman in the hull) were required.

The armor was SmK secure; all vital spots were to be protected by 13 mm.

As for performance, an average road speed of 25 to 30 km/h was required. Offroad speed was set at 20 km/h. For long-range performance, 150 km should be expected in six hours. The upgrade capability was to be 60% (31 degrees) on a stretch of at least one kilometer at a speed of 3 km/h. Further heightening of this performance was to be striven for. Climbing and wading ability were 600 mm each. With a ground clearance of at least 300 mm, a ditch-spanning ability of at least 1500 mm was required. The range with a full fuel load (on the road) was set at 150 km, but 200 km should be striven for. Full offroad capability in any terrain easily handled by infantry was called for. Ground pressure was 0.5 kh/sq.cm. As for communication, a radio set for telephone and

The "Light Tractor" made by Rheinmetall carried a 3.7 cm antitank gun in its turret. Only prototypes were built.

telegraph use was planned. The range of the telephone was to be 7 kilometers. Collective protection for safety from gas was required. A fixed device should produce fog for 20 minutes. Amphibian use was wanted, and should be made possible with the help of added swimming devices. The total weight was to be kept as low as possible, and should not exceed 7.5 tons.

On July 27, 1928, the Army Weapons Office informed WaPrüf 6 that the Daimler firm had specifically declined a development contract for the "Leichttraktor." WaPrüf was requested not to contact this firm any more. The memo T No. 691/28 Kdos II from the Army Weapons Office, of August 1, 1928, concerned the importation of war materials in A-Case. First of all, there were light and medium tanks from Britain, Sweden, and Czechoslovakia (R/R KH 50-Vollmer) involved.

In 1928-29 Rheinmetall carried on, along with development of the WD Tractor, development of the turret for the scout car and light tractor. One test piece was made and intended for the 8-wheel scout car. The turret held a 3/7 cm L/45 gun and a machine gun beside it. The traverse was 360 degrees, the elevation from -10 to +70 degrees.

The Rheinmetall firm also created the first armored self-propelled gun mounts, based on the Hanomag "WD" tractor chassis. The lighter vehicle was fitted with a 3.7 cm Tak L/35. 138 of them were to be built.

The situation with the "Grosstraktor" development, and the attitude of involved military offices toward test centers outside Germany, is shown in a letter that WaPrüf 6 sent to WaPrüf on 9/8/1928. It contained a series of considerations and suggestions:

"The purpose of the design and building of the 'Grosstraktor' was to create models from which, after thorough testing, a final model for later introduction and mass production should arise."

With the scope of the task, which was split into many parts, the goal could be reached only if the tests with the models now being worked on were undertaken in closest cooperation with all persons involved in WaPrüf, and the designers of the various firms. What with the intended testing in Russia, this was not at all possible. Personnel of the WaPrfw. could not be active in Russia for long periods, or repeatedly for shorter times, as was necessary. Aside from that, such long absence would limit the advancement of other important motorizing tasks considerably. It also seemed questionable whether all the firms, and individuals within them, were willing to work on testing in Russia. It was also questionable whether the people whose cooperation was necessary could always be in Russia together. The transport of vehicles to Russia was not

The 50 HP version of the WD "Schlepper" was used with the 7.7 cm cannon. This development was not taken further.

simple, and could bring many dangers of a domestic and international nature. Part testing, changes, and new designs of individual parts would be carried out only by the German manufacturers or design bureaus. Much transporting and considerable loss of time would result. Furthermore, Germany would give its newest and most secret technical advances to Russia without the assurance that sufficient use would result. Since all the travel and living costs—also for personnel of the various firms—had to be paid by the *Reichswehr* Ministry, testing in Russia produced extraordinarily high costs. How could one hope for the best results from the work to date and the extensive means in light of the strict domestic and international situation? Of the many problems in the development of the "Grosstraktor," one was basic, namely that of the running gear. All other tasks—gun turrets, machine-gun mountings, means of observation, FT equipment, gas security, air supply, armor, etc.—were worthless as long as the main problem—running gear—was not solved. Even outside Germany no usable solution had been found at this time, and it was stressed as the most vital task, ahead of all other problems. The other listed tasks were difficult, to be sure, but were not major problems like the running gear.

The formation of the tracked running gear caused almost insoluble problems at that time. Pictures show the first sprung running gear with steel-reinforced rubber tracks on a "Koen" tractor.

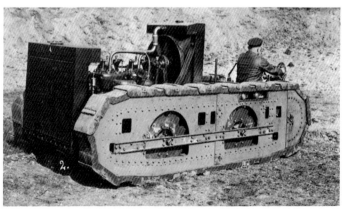

Another tracked running gear, this time unsprung, was that of the Type A tractor of the Wotan Works in Leipzig. This test vehicle, introduced in 1926, had a four-cylinder Daimler-Mercedes M 1574 engine that produced 100 HP.

Therefore, the following suggestion was made: Of the three models (two examples of each), one should be assembled at Unterlüss. Nothing should leave the assembly halls. They should serve only to test the entire construction, to the extent that this was possible on the premises. After a general impression has been gained, the three unassembled models should be disassembled as quickly as possible. Individual parts were of no danger, since the security at Unterlüss was regarded as good.

The assembly of the three other models should be halted at once. Everything was to be disassembled down to the chassis. The armor plate of each vehicle was to be cut up lengthwise. Only the running gear would remain, open at the top with engine and gearbox. Thus, the vehicle looked like a tracked tractor, such as were running for a long time at the WaPrüf 6 training command in Kummersdorf. With their rather large dimensions, the "Tractors" should be used then as heavy tractors for slabs. In addition, the vehicle offered a comparison with a fixed track-testing apparatus, in that it could be used as a mobile track-tester for tractor testing. In the vehicle testing company of WaPrüf 6 in Kummersdorf, the vehicles would attract less attention than in Unterlüss, where such activity was not constantly going on.

All the other parts should be dismantled in Unterlüss, or partly in Kummersdorf, too. The vehicles themselves should be packed and sent to Kummersdorf. Thus, the running gear could be tested best, most quickly, and cheaply. The further development of the other parts could go on unnoticed through individual tests. The amphibious tests were to be halted for a time, which could easily be taken in the bargain in comparison to solving the total problem. In the meantime, the small-model tests (1:10 scale) of the amphibian capability at the Technical Test Agency in Hamburg should be stopped. The best screw propeller shape had, after all, been determined. Two of these removable screws were intended to steer the vehicle in the water. Since WaPrüf accepted the suggestion to leave the vehicles in Germany, the first prototypes available since 1929 would be sent to Russia.

Rheinmetall began to develop the "Leichttraktor" under contract from the O.K.H. in 1929. In a contest with Krupp the firm built three test vehicles. They were laid out as tracked vehicles with Cletrac steering drive. A 3.7 cm l/45 Cannon and a machine gun were intended to go in the turret. The traverse was 360 degrees, the elevation ran from -10 to +30 degrees, and the vehicle weighed 8.5 tons. A 100 HP motor gave the vehicle a top speed of 35 km/h. A four-man crew was protected by 13 mm thick armor. The third vehicle was equipped as a self-propelled mount with a pivot mount.

In the production program of March 25, 1929, no. 1105/29 gKdos, the following costs were estimated:

Sf Tak	14	RM 280,000
Pivots and armor	138	RM 138,000
Track aggregates	70	RM 350,000
Armored vehicles	36	RM 3,600,000 4 test models
Large tractors	17	RM 3,400,000 6 test models
Light tractors	34	RM 1,700,000

Plus 24 Klkw (Dixi) at 80,000 Marks.

For the following years, the amounts in the budget for producing the In 6 (K) were set as follows:

1930	RM 2,727,730
1931	RM 3,488,811
1932	RM 4,212,732

The search for a usable interim solution for antitank defense became more and more urgent. A final report from the Weapons Office on the tests with an L.H.B. 50 HP tractor made it look questionable whether the tractor was suitable for the 3.7 cm Tak gun. Its body seemed very high, its width offered a too-large target, and its offroad capability would be insufficient. The Troop Office was requesting and hastening taking a stand on the performance of the 28 and 50 HP W.D. Tractors.

As of 1930, the first available "Kleintraktor" were subjected to thorough testing. On January 1, 1930, the tests were continued as follows: for the "Grosstraktor," all models were finished, and troop testing was begun. In 1930-31 RM 450,000, and in 1931-32 RM 1,000,000 were reserved for them. For the "Leichttraktor" chassis in the same time periods, RM 400,000 and RM 150,000 were available. Further development of these plans depended on whether the building of uniform vehicles was to be realized. For the "Kleintraktor" a conception was being worked on, for which RM 400,000 were set aside in 1931-32. Should the building of the "Grosstraktor" be dropped, sufficient means for the development of the "Kleintraktor" would be available in 1930. For the heavy self-propelled mounts (weapon carriers) for K-Flak and tank guns, models of which were being finished by Dürkopp and Krupp, RM 80,000 each had been reserved for 1930-31. An order was questionable, as only after the testing of all types was finished could a decision as to the choice of ideal types be made. For the development of the three uniform vehicles (light self-propelled mount, heavy self-propelled mount, and towing tractor) a further RM 300,000 had been scheduled.

Independently of the vehicle development, further tests that had great significance in the area of motorization were being made: synthetic motor oil was still in development, and its solution was of equal importance to that of synthetic gasoline. The solution of this task was in the most urgent interest of national defense. The location of new sources of domestic raw materials (oil shale) was not followed up because it was not economical. Yet the use of all sources was imperative to meet the national need for oils. An increased utilization of brown coal as a fuel was still being developed. This was an especially important matter on account of the favorable central location of the brown coal deposits in central Germany. The use of hard coal as a liquefied gas for fuel was tried.

The discussion of the development results on February 14, 1930, led to many basic standpoints: the subject of motors for special fuels, air-cooled motors, and fuels was recognized as important. Yet it could not be the job of military offices to apply considerable sums to it so as to lead the way for other interested agencies (post office, railroad, Ministry of Economics, business). The experience that the development was left to military agencies and the economy would then take over after favorable experiences could lead only to the conclusion that military offices, then as before, moved and inspired, but did not have the considerable means (RM 30,000 and 80,000) to put into the development.

For the towing tractor for antitank and infantry guns, it was determined that their use for antitank defense would cease. At that time a suitable small tractor was available, thus no further development was needed for the time being. As for the light self-propelled mount for antitank guns, the development of the Krupp tracked vehicle and the Horch wheeled/tracked vehicle were carried on, as large sums had already been devoted to them. It was also questionable whether the "Leichttraktor" chassis would be suitable for this purpose.

The acceptance of the development of the "Kleintraktor" was already arranged, thanks to a visit of the Chief of the Army Weapons Office in Kummersdorf. Regulations were still to be set up for its use as a) a reconnaissance vehicle, b) a weapon carrier, or c) a small towing tractor. Whether the vehicle was suitable as a small tank was questionable. Thus, a new designation (such as "armored reconnaissance car") was to be introduced, since the old one was misleading. The purchase of a British Carden-Lloyd chassis was urged, so that years of German development under prevailing circumstances would be unnecessary.

The testing of the "Grosstraktor" was carried out in 1930. A further decision was postponed. But no more vehicles of this type were built.

For the second production program (1933 to 1938), 37 "Light Tractors" and 20 "Small Tractors" were called for according to planning studies for the motorized troops on February 10, 1931. Special machine guns for installation in tanks were promoted, including three heavy machine guns for the "Grosstraktor," one for the self-propelled mount, and one MG for the "Leichttraktor" (planned Flz MG S-2-200). The request for a fully automatic 20 mm weapon was initiated for the "Kleintraktor" (tank destroyer) and for the makeshift tank. Thus, the schedule for 1931-32 was:

a) Grosstraktor
   1931 3 sMG installed
   1932 no changes
b) Leichttraktor 2 Süda MG (along with a 3.7 cm Tak)
c) Kleintraktor (Krupp tank destroyer)—a Rheinmetall 2 cm MG in a pivot mount is installed
   1931 wooden model finished
   1932 end of summer, tests with chassis finished, building wooden models and further development to follow.
d) Rheinmetall Sfl (Leichttraktor body, but open on top) is not developed further in 1931. One heavy MG installed along with the 3.7 cm Tak.

An order from the Gute-Hoffnung-Hütte was filled by their Swedish daughter firm, AB Landsverk. The picture shows the wheeled-tracked "Landsverk 30" vehicle, a design by Vollmer.

On the occasion of a discussion of the development program of In 6 with the leader of the WaPrüfw. on June 20, 1932, the following standpoints for the tractor development resulted:

"For the present building of the 'Kleintraktor,' changes are not undertaken. For which purposes this chassis would come into question later was to be determined in thorough tests. Until then, the already stated tactical-technical requirements will stand.

To what extent the further development of all tractors through design measures, or even new test models in the distant future, is to be foreseen, is to be decided at a conference on the basis of the summer's experiences in 'Kama' and the WaPrw. Vehicle tests at Kummersdorf. Even though the In 6 thoroughly recognizes the wishes for the return of the Krupp 'Grosstraktor' and further aggregates from the Daimler-Benz 'Grosstraktor,' this return cannot be brought about at this time for political reasons."

In 1930, Engineer Vollmer received an order from the Gute-Hoffnungs-Hütte in Sterkrade. For the daughter firm of AB Landsverk in Landskrona, Sweden, a light tank with wheel/track drive and a somewhat heavier full-track tank were to be developed. Heigl writes of it:

"One of the first test vehicles was the Type 'Landsverk 5,' in which the wheels were mounted on crank arms and moved up and down by means of a motor-driven spindle apparatus."

In the "Landsverk 30" model, which followed in 1931 on the basis of Vollmer's design (German designation RR 160), the transition from wheels to tracks could even be made while in motion. The full-tracked vehicle influenced by Vollmer was designated "Landsverk 10." As already in the Type 30, here too a 3.7 cm cannon was mounted in the turret. The fighting weight was 11 tons, and the crew numbered four men. Both vehicles were powered by 200 HP Maybach V-12 gasoline engines. The experience with these types, which were used mostly as civilian vehicles, naturally also benefited the contract-givers in Germany.

From 1930 on Rheinmetall was involved in the development of the so-called "Battalion Leader's Wagon," which arose from competition with Krupp. A test model was built by Rheinmetall, but the Krupp device was introduced after various changes, some of which depended on the Rheinmetall-Borsig design. The result was a tracked vehicle with Wilson steering gear and front drive. It was armed with a 7.5 cm gun and two machine guns. The vehicle weighed 18 tons; its motor produced 300 HP, and it reached a top speed of 35 km/h. The crew numbered five men. The front armor was 16 to 20 mm thick, and the side plates were 13 mm. This vehicle formed the basis for the later "*Panzerkampfwagen* IV."

The 7-5 cm cannon to be tested for use in self-propelled mounts weighed 440 kg, and had a length of 1950 mm (L/26). For the "Kleintraktor" development, a design by Krupp was on hand in the autumn of 1931. Through Russian cooperation it was possible to carry out the suggestion to obtain British Carden-Lloyd chassis. Two of them were bought for "Lama." In July 1932 Krupp introduced its Type "LaS" to the Weapons Office. Its disguised designation was "Agricultural Tractor" (LaS). This vehicle formed the basis for the later "*Panzerkampfwagen* I."

On February 3, 1932, the following armored tracked vehicles were on hand: two each of the Rheinmetall, Krupp, and Daimler-Benz "Grosstraktor," plus the "Leichttraktor" models by Krupp and Rheinmetall.

An improved version of the "LaS" vehicle was put into a five-piece 0-series production toward the end of 1932. It was introduced to In 6 in September 1933.

A speech at the *Reichswehr* Minister's office on November 14, 1933, on the subject of military improvements, resulted in the first three contracts being given to the Henschel firm in Kassel for 150 "LaS" vehicles, with the following directions:

"the finished tanks are, as long as they are not needed to set up tank formations, to go to the troop training sites. The production of a second series of tanks is to be promoted by WaPrüf at the right time."

Rheinmetall built prototypes of the so-called "Battalion Leader's Wagon," a forerunner of the later "*Panzerkampfwagen* IV," starting in 1930.

In the midsummer of 1933 "Kama" was shut down, and all the tanks located there were brought back to Germany. They were overhauled at the Daimler-Benz facility in Berlin-Marienfelde, and then sent to the Armored Firing School at Putlos, Holstein.

On the occasion of an earlier visit in "Kama" by General Lutz, suggestions for the building of the so-called "Neubaufahrzeug" had been given. Its development began in 1933 under contract from the O.K.H. to the Rheinmetall firm. In competition with Krupp, two vehicles were made with the Rheinmetall cast iron turret, while three vehicles were equipped with the armor-plate Krupp turret. Production stopped with these five vehicles. They were intended for tank attacks. Tracked vehicles resulted, equipped partly with Wilson and partly with Cletrac steering gear. They were armed with a gun turret and two machine-gun turrets. In the Rheinmetall version, one 7.5 cm L/23.5 gun, one 3.7 L/45 gun, and one machine gun were built into the turret. The guns were mounted in a shield, one over the other. The machine gun was placed in a ball mantlet beside the gun mount. The traverse was 360 degrees, the possible elevation from -10 to +22. In the Krupp version, the two guns were mounted side by side. All the guns weighed 370 kg.

The total weight of the vehicles was 23 tons. With the 290 HP BMW Va 6-cylinder aircraft engine, a top speed of 30 km/h was attained. Six-man crews were needed to operate these vehicles. Front armor was from 16 to 20 mm thick, side armor 13 mm. Vehicles 1 and 2 were used for training at Putlos, while Vehicles 3 and 5 were sent to Norway and used there during the war.

A light tank, Type LKA, planned by Krupp for other countries.

Only five of the "Neubaufahrzeug" were built by the Rheinmetall AG. Two were made of soft steel with the Rheinmetall turret. In this turret the 3.7 cm cannon was mounted over the 7.5 cm primary weapon.

The other three vehicles were made of armor steel, and fitted with the Krupp turret. Here the two cannons were mounted side by side.

Above: These three tanks were used in Norway during the war. The picture shows the small front and rear turrets, each armed with an MG 13. Below: Another MG 13 was mounted in a firing flap. The driver's seat is near the front turret.

The production of a second series of tanks, urged in 1933, resulted in the first "LaS 100" prototype, later the "*Panzerkampfwagen* II" of the Augsburg-Nürnberg Maschinenfabrik, in 1935. At Rheinmetall, under contract from the O.K.H., the development of a "3.7 cm Self-propelled Mount L/45" began in 1934. The "Leichttraktor" was fitted with a new turret. The armament consisted of a 3.7 cm Pak L/45 and a light machine gun. The vehicle weighed 8.5 tons, and the top speed was 35 km/h. The powerplant produced 100 HP. The three-man crew was protected by 13 mm armor. The vehicle was not sent to the troops.

In 1934 Rheinmetall was also busy with the development of a tank turret for the "ZW" vehicle, the later "*Panzerkampfwagen* III." In cooperation with Daimler-Benz, which developed the vehicle, a test model was built for the Daimler-Benz platoon leader's car The turret was armed with a 3.7 cm gun in a shield, and two machine guns in a special shield next to that of the cannon. Another MG was planned for the bow of the vehicle. The traverse field was 360 degrees, and the elevation ranged from -10 to +20 degrees.

The drive wheels were in the back, an arrangement never again used by the German *Wehrmacht's* later production vehicles.

For the vehicle disguised as "LaS," the following numbers were produced by December 31, 1934: 150 for 7.5 million RM and 300 bodies for 208 million RM. From December 1934 to March 1935, a monthly production of 40 to 75 tanks was planned. Since only five of the E.MG. 33 (later MG 34) were available, the Dreyse MG in a cast iron shield had to be used, since great numbers of machine guns were needed for installation in tanks. The turretless vehicles turned over to the troops for training in the first years had the official designation of "Krupp-Traktor."

It is noteworthy that, despite the few tangible developments, which obviously went in different directions (Grosstraktor, Neubaufahrzeug), the concept for later was clearly defined already in 1930. When one regards the transitional solutions (Leichttraktor, LaS, and LaS 100) in this light, one can see in the development of the "BW" and "ZW" vehicles the guiding influence of Heinz Guderian. Through the clever division of the crew's tasks, extremely effective fighting instruments were made for the awaited armored divisions. Germany, free from any traditional limits, had decided on a revolutionary way to wage war.

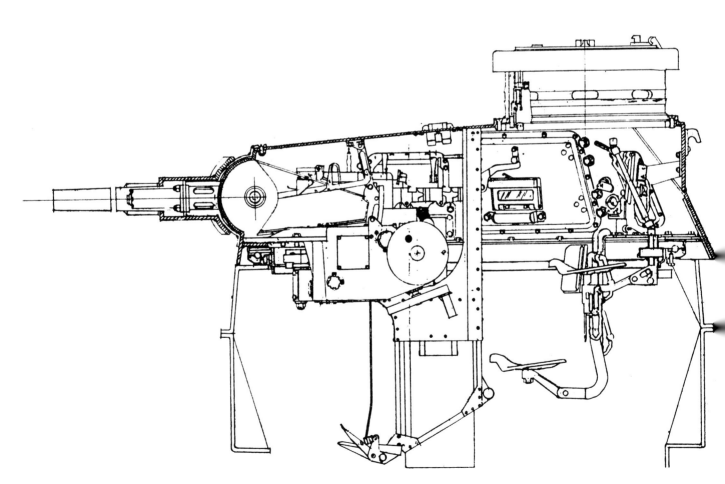

The Rheinmetall firm worked from 1934 on to develop an armored turret for the "Zugführerwagen," the later "Panzerkampfwagen III."

Panzerkampfwagen I and Variations

After the MAN, Krupp, Henschel, Daimler-Benz, and Rheinmetall-Borsig had presented designs for a light tank of the five-ton class, the Army Weapons Office chose the following builders for the final development:

Friedrich Krupp AG, Essen, for the chassis.
Daimler-Benz AG in Berlin-Marienfelde for the body.

The firms of Henschel & Sohn in Kassel and the Krupp-Gruson AG in Magedburg were the main ones chosen for production, with MAN of Nürnberg and Wegmann AG of Kassel added later.

Henschel had finished the first pre-series vehicles in December 1933; their first test drive took place on February 3, 1934. At Krupp-Gruson, the first three prototypes were already finished at the end of 1933. The production designation of these vehicles was "I A LaS Krupp." The vehicle as presented in its final form was a full-track type with a two-man crew, and two MG 13 guns mounted in a turret.

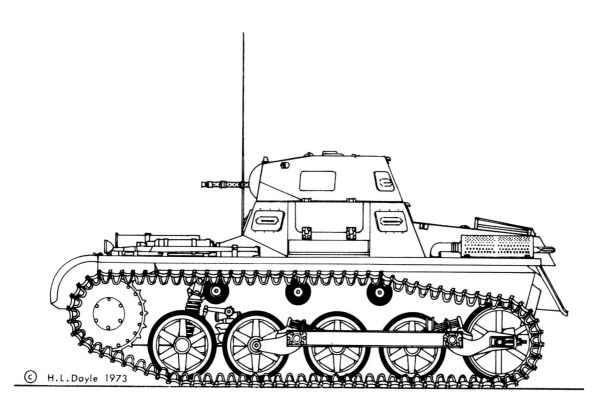

"*Panzerkampfwagen* I" (MG), Type A (Sd.Kfz. 101).

The horseshoe-shaped turret had slightly angled walls, and had room only for the gunner. The elevation and traverse were set by handwheels. Four visors were set evenly around the turret. A large entry hatch was built into the roof. The driver got into the vehicle through a hatch on the vehicle's left side. He sat to the left, facing forward. His view was limited, since he had only two visors looking forward and to the left. The gunner in the turret was also the tank commander.

The air-cooled 60 HP Krupp "M 305" four-cylinder motor had a bore of 92 mm, a stroke of 130 mm, and a top rate of 2500 RPM. It was attached to the central wall, along with the intermediate gears, and to the rear by a sprung bracket.

The motor was cooled by making the cylinder walls with numerous ribs close to each other, and having a strong stream of air blow over them. The valves were cooled by a special direction of the airstream. The needed air was pulled from outside through the oil cooler, which was between the two fuel tanks, by a fan mounted on the crankshaft. The warm air was led by guiding panels over the fuel tanks to the air exits, which were on both sides of the vehicle's rear.

The air-cooled "M 305" gasoline engine made by Krupp, and used in "*Panzerkampfwagen* I," type A. Motor, seen from right:

1. Oil dipstick
2. Drain plug
3. Splinter Filter
4. Oil filter
5. Fuel pump
6. Distributor
7. Coil
8. Exhaust manifold
9. Generator
10. Clutch shaft
11. Crankcase
12. Oil line to distributor-starter casing
13. Oilpan
14. Right exhaust pipe
15. Oil line to rocker
16. Protective cover
17. Cylinder head
18. Valve lever housing

The fuel was carried in two 72-liter tanks and fed by pumps. The power flow went through an intermediate gear, an intermediate shaft, and a dry two-plate main clutch to the ZF Aphon "FG 35" five-speed gearbox. From there a bevel drive operated the "clutch steering gear."

This steering gear carried the turning moment to both sides and to the tracks. It was flanged on the shift drive, and consisted of a bevel drive and two steering clutches, mounted on either side of the bevel wheel for the differential drive. Each clutch had a band brake on its exit side. The steering clutches uncoupled the connection of the motor to the tracks by pulling the steering levers.

If the brake on the steering clutch was activated by further pulling of the steering levers the track was stopped, and thus the vehicle was steered.

The two intermediate gears were flanged onto the inside of the hull walls. In each gear there were a pinion and a toothed wheel that served to reduce the number of revolutions between the differential and the drive wheel.

The running gear on each side consisted of a drive wheel, the road wheels (size 530 x 80), the leading wheel, and three jack rollers (size 190 x 85). A track was stretched around each set. The drive wheel had a changeable geared ring.

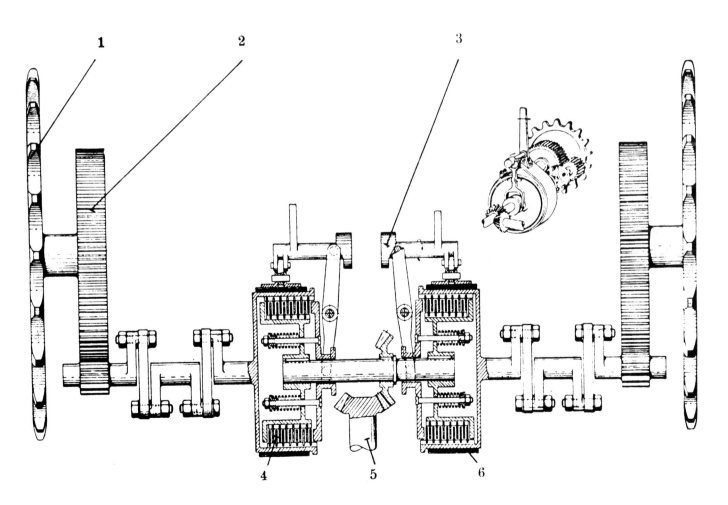

The clutch steering drive of "*Panzerkampfwagen* 1," simplified: 1. Drive wheels; 2. Intermediate gears; 3. Clutch cams; 4. Clutch; 5. Bevel gear; 6. Brake band.

The front road wheel was suspended on a crank arm attached to the front tube axle. It was sprung by a coil spring, secured from bending, which was sprung below against the crank arm and above against a ball-pan bearing attached to the hull. Damping the spring's swinging was done by Boge shock absorbers. The movement of the crank arm was limited above by rubber buffers in the spring guide, and below by the attachment cast into the crank arm and its bearing.

The second and third road wheels were suspended in a running-gear lever that was centrally mounted on the second tube axle and able to turn. Similarly, the fourth road wheel and the leading wheel were suspended on a running-gear lever that was mounted on the third tube axle and able to turn. The second and third axles were linked together by a U-rail.

The running-gear levers of the second and third road wheels consisted of a cast steel lever, the spring packets, and two pairs of

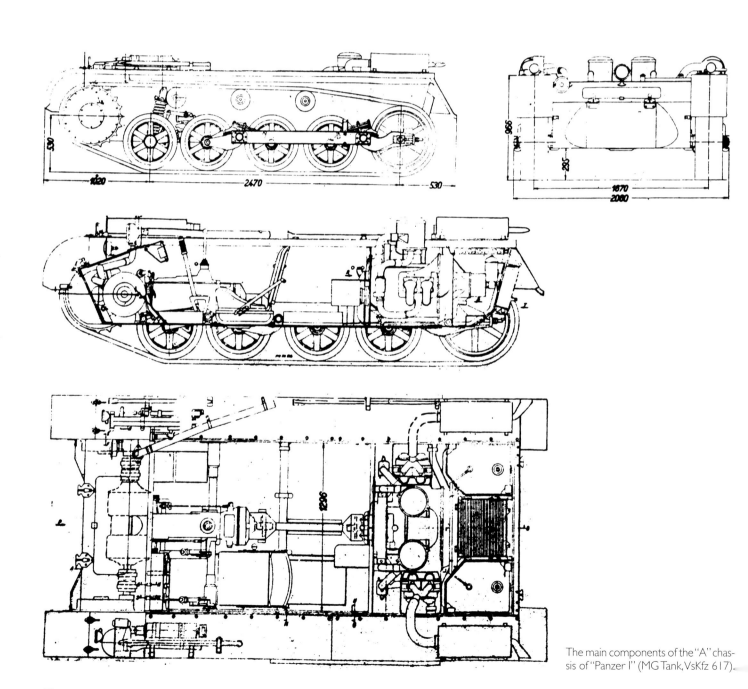

The main components of the "A" chassis of "Panzer I" (MG Tank, VsKfz 617).

For the first version of the "*Panzerkampfwagen* I" a Carden-Lloyd chassis was used, which was recognizable by its low-lying leading wheel.

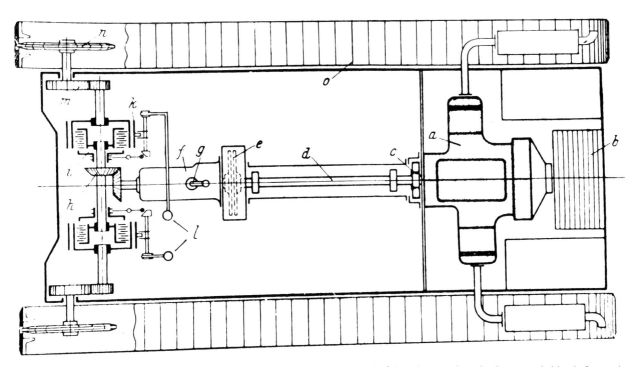

The complete layout of the "Panzer I" chassis in a schematic drawing: a. Motor; b. Oil cooler; c. main reduction gear; d. driveshaft; e. main clutch; f. gearbox; g. Shift lever; h. Crown wheel and bevel gears; i. Steering clutches; k. Clutch brake; l. Steering lever; m. intermediate gear; n. Drive wheels; o. Tracks.

coil springs lying in the cranks. The levers for the fourth road wheel and the leading wheel were similar to those for the second and third road wheels.

The road wheels consisted of a light metal wheel with cast sockets and vulcanized rubber tires. The tracks consisted of individual ungreased links 280 mm wide, linked together with bolts. The tracks were tensed by adjusting the leading wheels. The low-lying leading wheels on this vehicle were noteworthy.

The 5.4-ton vehicle had a top speed of 40 km/h. The armor plate, 13 mm all around, made it safe from S.m.K. fire. During the war 15 mm armor plates were often added.

In some of these vehicles the air-cooled Diesel version of the Krupp Type "M 601" opposed engine was installed experimentally. With almost the same dimensions, this powerplant produced some 45 HP at 200 rpm. This performance turned out to be insufficient. Although some vehicles of the B version were still fitted with it, the tests were soon halted. The vehicles were easy to recognize by the mufflers mounted outside on the side of the engine compartment. It is worth mentioning here that until almost 1940 (Tatra "103" Diesel engine"), no further tests were made to find an air-cooled Diesel engine for tanks.

A look at "*Panzerkampfwagen* I" (MG), Type A (Sd.Kfz. 101).

Opposite: These vehicles were displayed in great numbers from 1933 on, including at political events. This display was part of a harvest festival in Bückeburg.

In some "Panzer I" vehicles the air-cooled Krupp "M 601" Diesel engine was installed experimentally. Typical features of this version were the enlarged mufflers and the hull shortened at the rear.

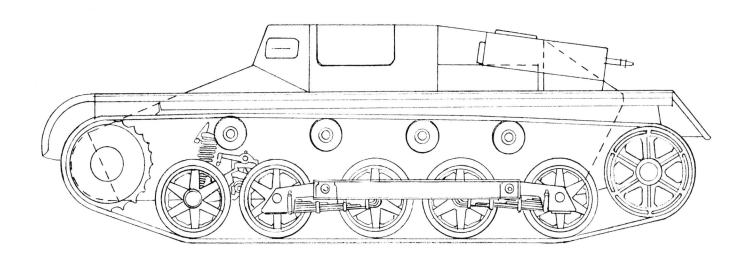

Series production of the "I A LaS Krupp" vehicles began at the Krupp-Gruson AG in the spring of 1934, while the Henschel works began production in July 1934. By 1937 Krupp had produced about 750 of these tanks (Types A and B). Henschel built 349 "Panzer I," and the Augsburg-Nürnberg Maschinenfabrik built 128 "I A LaS Krupp" vehicles. The first contracts for 150 tanks were followed by others for almost 1000 of them.

The finished tanks were designated "*Panzerkampfwagen* I" (MG) (Sd.Kfz. 101) Type A and sent to the troops, forming the nucleus of the vehicles of the new armored units.

In all, 477 Type A tanks (chassis numbers 10 001 to 10 477) were delivered. The vehicles tested by the troops very soon showed that the powerplant only conditionally met the established requirements.

Further development could not be avoided. In 1935 the improved version of Panzer I, again supervised by Krupp, appeared, designated "I B LaS May" (chassis numbers 10 478 to 15 000 and 15 201 to 16 500). Daimler-Benz was responsible for the body changes. The main differences applied to the running gear and engine compartment. Now the water-cooled Maybach "NL 38" motor, with 90 mm bore, 100 mm stroke, 3.8 liter displacement, producing 100 HP as 3000 rpm, was installed.

The establishment of new armored units was done quickly after Hitler came to power. The "*Panzerkampfwagen* I" was the nucleus of the future armored divisions.

During the war a few "Panzer I" tanks were rebuilt into flamethrower tanks by the troops. This picture shows such a tank in action in North Africa.

To transport the "Panzer I" for long distances, a number of heavy trucks were obtained by the *Wehrmacht*. This picture shows a Büssing-NAG Type "654" equipped with four-wheel drive, cable winch, and removable ramps.

Three-axle trucks made by Büssing-NAG and Faun were also used to transport light tanks. The pictures show such vehicles of Fan Type "L 900 D 567."

This view of Type A "Panzer I" shows the changed air intake to the engine compartment, and the changed arrangement of the exhaust system.

On the other hand, Type B of "Panzer I" has a lengthened engine compartment, now holding the six-cylinder Maybach "NL 38" motor. The different arrangement of the exhaust system is easy to see.

Comparison of Types A and B.

47

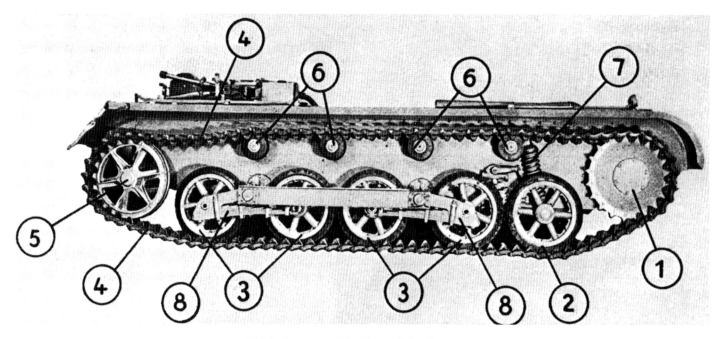

1. Drive wheel
2. Front shock-absorber wheel
3. Road wheels

The further parts of the "Panzer I" Type B chassis:
4. Track
5. Leading wheel
6. Jack rollers

7. Shock absorber
8. chassis spring

The fuel was carried in two tanks, the front one holding 84, the rear one 62 liters. Fuel was fed by pumps. The size of this powerplant required a lengthening of the engine compartment, and thus of the hull. This could only be done by adding a pair of road wheels to the running gear. The running gear on each side now consisted of the drive wheel, four road wheels, the leading wheel, and the four jack rollers. In its layout it agreed with that of Type A. The leading wheels, though, were raised and mounted on crank arms in boxes welded at the rear. The tracks were tensed by swinging the crank arms. So as not to limit the steering, and thus the mobility of the vehicle, the length of track on the ground was not increased. The fighting weight of the tank grew to about six tons.

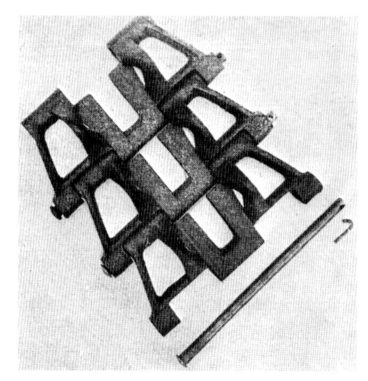

The track of the "Panzer I"; shown are three links, attached by ungreased bolts.

The chassis of the "Panzer I," looking forward:
1. Driveshaft cover
2. Clutch housing
3. Gearbox
4. Shift lever
5. Steering brake housing
6. Ventilator motor
7. Air shaft (exit)
8. Intermediate gear
9. Steering lever
10. Instrument panel
11. Clutch pedal
12. Brake pedal
13. Accelerator pedal
14. Driver's seat (adjustable)
15. Starter battery
16. Tube axle cover

The chassis of the "Panzer I," looking backward:
1. Starter motor
2. Radiator
3. Radiator vent
4. Exhaust pipe
5. Magneto
6. Air filter
7. Accelerator
8. Driveshaft cover
9. Tube axle
10. Fuel filler
11. Fuel tank
12. Driver's seat

49

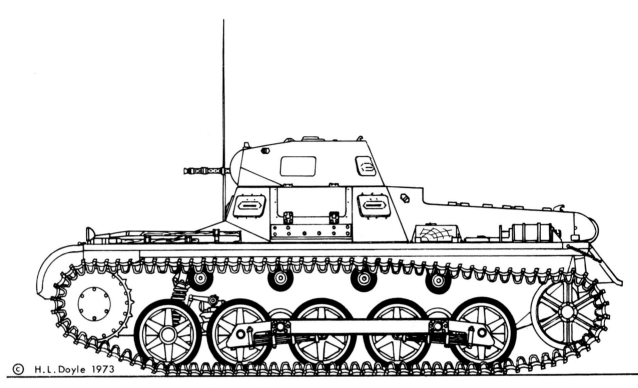

Panzerkampfwagen I (MG), Type B (Sd.Kfz. 101).

The improved chassis of the "B" version had a higher leading wheel.

The entire length increased from 4.02 to 4.42 meters. Otherwise, the vehicle remained unchanged in technical and tactical terms. Only an improved ZF Aphon "FG 31" gearbox was used.

The vehicles that reached the troops toward the end of 1935 were officially designated "*Panzerkampfwagen* I" (MG) (Sd.Kfz. 101), Type B. The price of one (without weapons) was 50,000 Reichsmark.

Henschel und Sohn produced most of these vehicles between 1935 and 1937, while MAN (75 tanks in 1936-37) and Wegmann (1936-39) were also involved in their assembly.

The main supplier of the bodies was the Deutsche Edelstahlwerke AG in Hannover-Linden, which delivered, among others, the following parts for the Panzer I program:

Year	Hulls	Upper Bodies	Turrets
1933	31	—	—
1934	337	54	54
1935	811	851	851
1936	574	565	557
1937	114	255	31
1938	—	22	—

More exact production figures are not available, but about 1900 "Panzer I" tanks were built. The "Panzer I" models saw their first service in the Spanish Civil War. Some of the tanks left there were later fitted with bigger turrets, holding a 20 mm cannon, by the Spanish Army.

On September 1, 1939, 1445 Panzer I were on hand, and when the French campaign began in 1940, there were still 523 of them in the motor parks of the armored divisions. According to figures from July 1, 1941, there were 842 tanks on hand in the Army.

The track-tightening apparatus of the leading wheels was easy to reach. The chassis of "*Panzerkampfwagen* I," Type B is shown at an angle from the rear.

A look at the "*Panzerkampfwagen* I" (MG) Type B (Sd.Kfz. 101).

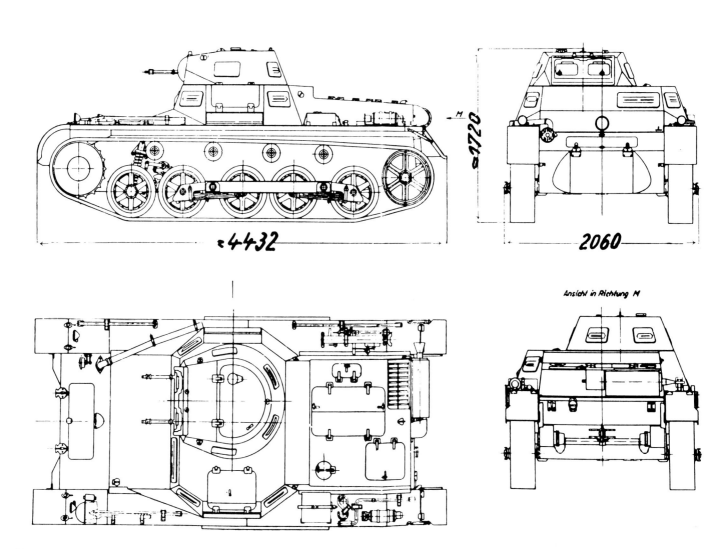

"*Panzerkampfwagen* I" in action in Poland in 1939. The white German crosses were used only during this campaign.

"Panzer I" also took part in the occupation of Denmark. Further action raised questions about the field usefulness of these tanks.

Above: The "C" version of "Panzer I"—not built in large series—had upgraded armament, and was designed for higher speeds. Below: This "VK 601" had a box running gear, suspended on transverse torsion bars. The fog-laying apparatus attached to the right front fender is also interesting. In VK 601, as in the Panzer II Types D and E, the teeth of the drive wheels were changed, and greased tractor tracks with needle bearings were used experimentally.

A look at the Maybach pre-selector gearbox of "VG 15319." No place for the radioman was planned in the bow of this vehicle.

The radio set (here Fu Spr a) was housed in the fighting compartment.

The driver's seat of "VK 601" with its side visor. The tank had a steering wheel instead of the usual lever.

The Panzer I tanks of both types that saw service in North Africa had been fitted with changed ventilation and additional air filters. The vehicles thus equipped added the suffix "Tp" (tropic) to their designation.

Heinz Guderian wrote in his book *Memoirs of a Soldier*: "It (the "Panzer I" vehicle) could be made ready for front service with this limitation by 1934, and serve at least as a training tank until the battle tans were finished. The introduction of this device was thus ordered under the name of 'Panzer I.' Nobody thought in 1932 that we would have to go at the enemy one day with these little practice tanks...." The action in Poland, France, and North Africa also showed very quickly that neither the firepower, nor the armor of these vehicles allowed them to fight against enemy tanks. Since the supply of larger vehicles was at first more or less assured, Panzer I tanks were mustered out, slowly at first, then more and more quickly, and by the end of 1941 they had almost completely disappeared as battle tanks.

This "VK 1801" can be regarded as a typical example of infantry-support tanks. This heavily armored Type F of "Panzer I" was, like "VK 601," built by Kraus-Maffei.

On September 15, 1939, the AHA/Ag L/In 6 gave a contract to the Weapons Office for a light tank for reconnaissance purposes, which was also to be usable by airborne troops. The development firm for the chassis of this project was Kraus-Maffei of Munich, while Daimler-Benz in Berlin-Marienfelde was to provide the body and turret. With a total weight of the tank at about eight tons, armor of 10 to 30 mm was planned. A Maybach "HL 45 p" gasoline engine producing 150 HP was to be installed, giving the vehicle a top speed of 65 km/h. The crew of two men had a 7.9 mm EW 141 and an MG 34 available in the turret.

This "*Panzerkampfwagen* n.A. VK. 601," of which a test series of 46 tanks was contracted for, was also known as "Panzer I (Type C)(VK. 601)."

The further development of *Panzerkampfwagen* I, with emphasis on heavier armor, led to a contract of December 22, 1939, for a 0 series of thirty "*Panzerkampfwagen* VK. 1801." This contract also went to the Krauss-Maffei and Daimler-Benz firms.

The arrangement of the large circular entry hatches on both sides of the hull was unusual. The gross weight, increased by the heavy armor, required wider tracks.

Open chassis were used for driving school purposes. Chassis of both versions were used.

During the war, tank driving training was taken over by the NSKK (National Socialist Driving Corps). Here the controls of "Panzer I" are being explained.

Front armor of 80 mm was required, raising the total weight to 18 to 19 tons. As in the VK. 601, the Maybach "HL 45" motor was to be installed. While the "VK. 601" used the Maybach "VG 15319" pre-selector gearbox, the "VK 1801" was to use the "SSG 47" four-speed gearbox. Two MG 34 were mounted in the turret. The crew numbered two men. The first chassis was finished on June 17, 1940, and the first turret was ready. The series began at the end of 1940, and was delivered until 1942. A subsequent contract for 100 of these vehicles was withdrawn. The Kraus-Maffei firm received a test contract in March 1940 for the installation of radio sets (Fu 2 and on-board speakers) in VK. 1801, but this test was given up. This development was designated "*Panzerkampfwagen* 1 n.A. verstärkt" or, according to D 65-/33, as "*Panzerkampfwagen* I (Ausf. F)(VK 1801)." Chassis numbers began with 150 301.

Open-hulled "Panzer I" models were used for driving school purposes, especially when the NSKK took over driving instruction during the war. A variant of the "A" type of Panzer I was the "*Panzerkampfwagen* I (A) Ammunition Tractor." This vehicle, developed by Krupp and Daimler-Benz, was used by the fast troops as a supply vehicle. It had a weight of five tons, a two-man crew, and armor 15 mm thick on the front and 13 mm on the sides. These Sd.Kfz. 111 were only 1400 mm high, and supplied armored troops with ammunition and, when necessary, fuel. In 1940 there were 51 of them on hand. They were unarmed.

Versions of the earlier tanks were made by removing the turrets, especially by the vehicle repair groups with armored vehicles. Both versions were used. They were designated "Instandsetzungs-kraftwagen I." The armored engineer companies, according to K.St.N. 716 of March 6, 1940, were each given two companies of "Pionier-Kampfwagen I." These vehicles had no turrets, but rather special bodies to house engineer equipment.

The numerous self-propelled mount developments began with mustered-out Panzer I tanks. By using the "B" chassis, a tank

"Panzer I" chassis of the NSKK are driven offroad at the driving school.

The Altmärkische Kettenfabrik built a large number of "B" chassis into tank destroyers. This was the beginning of a long development of similar vehicles. The Czech 4.7 cm antitank cannon was used as their armament.

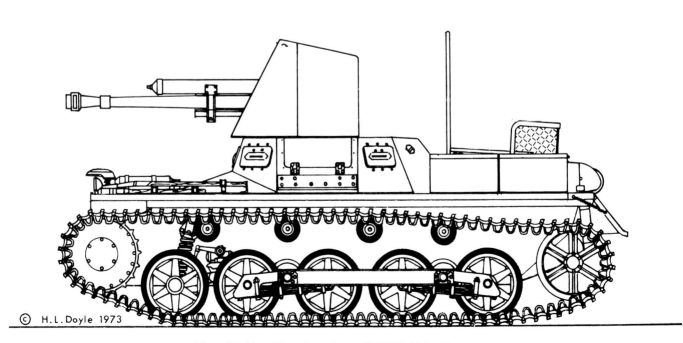

4.7 cm Pak (t) on "*Panzerkampfwagen* I" (Sd.Kfz. 101) without turret.

ALKETT also mounted a number of heavy infantry guns on "Panzer I B" chassis.

The rebuilding of these vehicles took place at the Skoda works in Pilsen. The original chassis were fully dismantled, reinforced, and rebuilt. The same was done to Czech Panzer 35 (t) tanks.

The chassis were overloaded, the elevation impossible for tactical use. Yet these vehicles, especially in France, made important contributions to the support of motorized infantry.

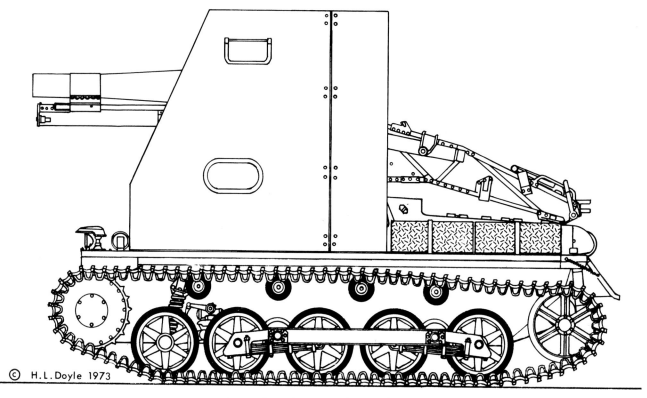

15-om-sLG on "Panzerkampfwagen I," Type B.

destroyer was created by Alkett of Berlin, and a one-time series of 132 vehicles was delivered. The designation was "4/7 cm Pak (t) on *Panzerkampfwagen* I (Sd.Kfz. 101) ohne Turm."

The delivery numbers were planned as follows: March 1940: 40; April 1940: 50; and May 1940: 42. The Daimler-Benz and Büssing-NAG firms, both in Berlin, were to do the rebuilding. The 6.4-ton rebuilt vehicles had three-men crews. A 4.7-ton Pak (t) was installed behind a 14.5 mm armor shield, with a traversing field of 10 degrees to each side, and an elevation of +17.5 to -8 degrees. The body was open on top and to the rear, and could carry 86 rounds of ammunition. The weapon was of Czech origin, and had a barrel length of 2040 mm (L/43.4). The maximum shot range (at 25-degree elevation) was 6000 meters. The firing height was 1720 mm. These vehicles were used in North Africa and Russia.

Turretless "Panzer I" vehicles were used in great numbers by the repair units of the armored troops.

The armored engineers also received a number of "Panzer I" vehicles fitted with open bodies. Both A and B chassis were used.

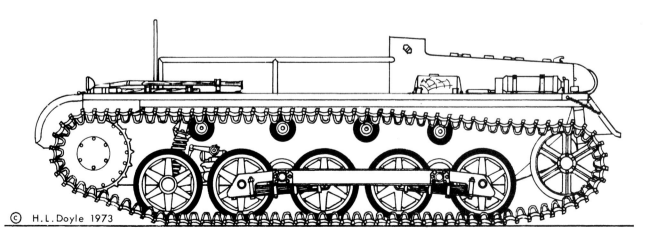

Instandsetzungskraftwagen I.

Also built by the Altmärkische Kettenfabrik GmbH was a self-propelled artillery mount that was planned as a support weapon for the armored grenadiers. Here a complete 15 cm sIG 33 was loaded onto a "Panzer I" chassis, and protected on three sides with 10 mm armor plate. The fighting compartment remained open on top and in back. This version, as original as it seemed, created a vehicle with the impossible height of 3.35 meters. The chassis was also fully overloaded, since the weapon alone when ready to fire weighed 1.75 tons. With a four-man crew the fighting weight of the vehicle was about 8.5 tons. These vehicles were used at the beginning of the war, and supported penetrated tank units effectively with the effect of heavy fire. The road conditions in Russia brought a fast end to their service. Those used by the troops were designated "15 cm sIG 33 auf *Panzerkampfwagen* I Ausf. B," and in individual cases also "Geschützwagen (Gw) I für 15 cm sIG 33." There were 38 of them in 1940-41.

Krupp and Daimler-Benz developed these armored supply vehicles on the chassis of the Type A "Panzer I." They were designated "*Panzerkampfwagen* I (A Munitionsschlepper (Sd.Kfz. 111)."

The first version of the "Ladungsleger" had a sliding ramp over the engine cover, on which an explosive charge was placed. When the vehicle had backed to the target to be destroyed, the charge was released and then ignited by a time fuse.

The rear view of the vehicle shows details of the sliding ramp.

The improved version of the "Ladungsleger I" had an extendable swinging arm with which the charge could be unloaded to the front or back.

The rear view of the "Ladungsleger I" shows the otherwise unchanged "Panzer I" vehicle with the unloading apparatus built on.

Many attempts to create radio-controlled tanks and so-called charge carriers used the chassis of the obsolete Panzer I. The Waggonfabrik Talbot in Aachen took a major part in these scarcely known attempts. At first a sliding ramp was attached over the engine compartment, with a box-shaped explosive charge resting on its rails. This charge, equipped with a time fuse, was driven up to the target backward and released from inside the tank. The time fuse allowed the vehicle to depart before the explosion. Since these methods were comparatively primitive, Talbot, under contract as of May 9, 1940, created the "Ladungsleger I." Here an apparatus allowed a 75-kg explosive charge to be unloaded from a Panzer I by a swinging and extendable arm. The arm, attached at the end of the fighting compartment, was two meters long at rest, and could be extended to 2.75 meters. The explosive charge could also be unloaded to the front. Only test models were made.

Some Panzer I were also used as flamethrowing tanks during the campaign in North Africa. Obviously, they were rebuildings made by the troops themselves. In several versions of the A vehicles, the right machine gun was removed and a canister Flamethrower 40 installed. The left MG was retained. The burning fluid and air-pressure tank were installed inside the tank. Eight one-second bursts of fire with a range of up to 25 meters were possible.

A "Small *Panzerbefehlswagen*" (Sd.Kfz. 265 on the "Panzer I" Type A chassis).

The efforts to supply the armored units with armored command vehicles led to the creation of the so-called "Panzerbefehlswagen." The lighter version used the chassis of Panzer I. The development firm for the chassis was, as before, the Krupp firm. The new type of body was developed by the Daimler-Benz AG. As of 1936, the first versions of the "kleiner Panzerbefehlswagen (Sd.Kfz. 265)" appeared. The body, on the normal "Panzer I" chassis, was considerably expanded and laid out in box form. The front body plate consisted of one piece of armor plate. Plates 14.5 mm thick were used all around the body. The body was attached to the hull with screw connectors. On the right side of the roof was a narrow armored cupola from which to observe the battlefield.

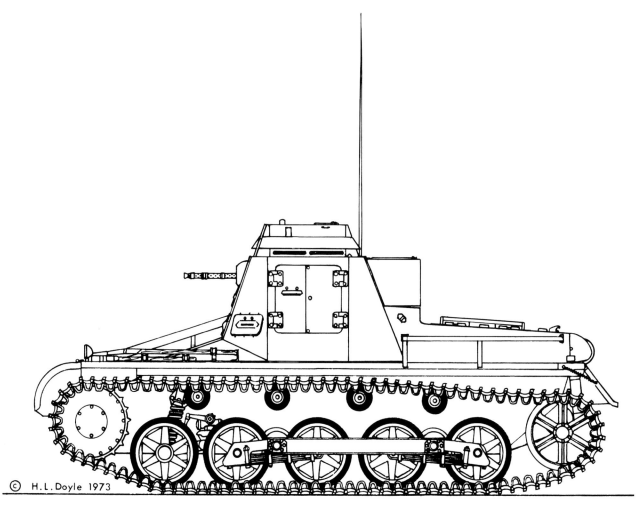

Kleiner *Panzerbefehlswagen* (Sd.Kfz. 265, Type B).

These vehicles saw service in the occupation of Austria in 1938. The picture shows the rear view of the small armored command vehicle.

Another vehicle in the Polish campaign of 1939. The MG 34 in a ball mantlet is easy to see.

Access to the fighting compartment was through a two-part door on the left side of the vehicle. The roof cupola was likewise provided with an exit hatch. The only armament was an MG 34 in a ball mantlet, built into the upper right side of the front plate. Three crewmen were foreseen: a driver, a radioman who used a "Fu 6" and an "Fu 2" set, and a commander. The first version, with the DB designation "1 KI A," used the A version of the Panzer I; the following production models "2 KI B" and "3 KI B" used the B type. Chassis numbers ran from 15 001 to 15 200. In all, 200 of these vehicles were built. The total weight of the vehicle was 5.88 tons. The bodies were built by the Deutsche Edelstahlwerke AG in Hannover-Linden. At the beginning of the French campaign in 1940, 96 of these vehicles were available to the attacking troops.

The idea of developing two-man tanks turned up again surprisingly toward the end of the war. According to available but unreliable information, the Weserhütte AG in Bad Oeynhausen worked on the design of a VK. 301, while Büssing-NAG developed plans for a VK. 501. Both designs were dropped by the end of the war.

The last versions of the small armored command vehicle used the "B" chassis. Note the large entry hatch of this vehicle used for command activities.

Panzerkampfwagen II and Variations

About the background of the production of a second series of tanks that was called for in 1933, *Generaloberst* Heinz Guderian says as follows in his book, *Memoirs of a Soldier*:

"...since the production of the planned main types (Panzer III and IV) was taking longer than had originally been hoped, General Lutz decided on a further interim solution, the Panzer II, armed with a 2 cm machine cannon and a machine gun..."

The Weapons Office, occupied with its development, gave appropriate contracts for an armored vehicle of the 10-ton class to the firms of Friedrich Krupp AG in Essen, Henschel und Sohn AG in Kassel, and the Maschinenfabrik Augsburg-Nürnberg AG in Nürnberg. The first prototypes were shown to the representatives of the In 6 in the spring of 1935.

Under the Army Weapons Office's disguised designation of "LaS 100" (*Landwirtschaftlicher Schlepper* 100), these prototypes were tested thoroughly, even though the urgent need of these vehicles did not allow a constructive development before series production began. After the first tests with the MAN prototype had gone satisfactorily, the Weapons Office decided to give contracts to the eventual developmental firms:

Maschinenfabrik Augsburg Nürnberg AG, *Nürnberg* factory, for the chassis.
Daimler-Benz AG, Berlin-Marienfelde factory, for the body.

After series production began, the following firms were involved in the production: FAMO in Breslau (1935-43), Wegmann in Kassel (1935-41), and MIAG in Braunschweig (1936-40). The first production chassis finished by MAN beginning in August 1936 were designated "a1" (chassis numbers 20 001 to 20 010), and were issued to the troops as Type "1/LaS 100"; they were officially known during 1936 as "*Panzerkampfwagen* II (2 cm) (Sd.Kfz. 121)."

These 7.6-ton vehicles were armored so that in deviations of the vehicle up to 30 degrees from the horizontal going forward and 15 degrees sideways, cannon shots could not penetrate from any range. The armored hull consisted of armor plates welded together, and was reinforced in front and back by angle irons welded to the bottom. The axle pipes on which the pairs of road wheels were mounted were attached to the sidewalls of the hull.

The armored hull served as a chassis frame, and held the motor and the powertrain components, as well as the crew.

A Maybach 6-cylinder "HL 57 TR" engine displacing 5.7 liters was installed. With 100 mm bore and 120 mm stroke, it produced 130 HP at 2600 rpm. The radiator was formed so that, with an outside temperature of 35 degrees, even with lasting full-load running, the water temperature did not exceed 95 degrees.

The gearbox itself was a geared type with a geared clutch. It had six forward speeds and one reverse. It handled torque up to 45 m/kg. An intermediate shaft carried the power of the motor to the main clutch, which was a dry two-plate type. The steering gear consisted of two planetary drives. The first served as a clutch to

The Krupp suggestion for the "Panzer II" series was the "LKA 2" vehicle, a further development of the "Panzer I" prototype. A 2 cm primary weapon was planned.

Henschel took part in this project, and produced a test chassis for a 10-ton vehicle.

release the power from the braked track while steering; the other was a reduction gear, and as such it was linked with a brake disc to brake the track in steering and stop the vehicle.

The six-wheel leaf spring running gear had six road wheels on each side, united in three pairs, and mounted in rockers on the hull. One roller of the pair was mounted firmly in the rocker, the other in the leaf spring. The springs of the individual wheel rockers were synchronized with the pressure coming to them. The chassis ran on two ungreased tracks. The tracks were driven by drive wheels mounted at the front and linked with the steering gears. The tracks were carried by the road wheels, which ran between the track teeth and transferred the vehicle weight to the tracks. At the rear end of the chassis, the tracks were directed forward over the leading wheels. The leading wheels of Type "a1" were cast of Silumin, and had rubber coverings. From the leading wheels, the tracks ran over three jack rollers, which supported the returning part of the track. The track was tightened by adjusting the leading wheel. The arrangement of the drives next to the driver created a driver's area that was drawn back sharply into an angle on the right side to make the installation of a visor flap possible. The bow of the hull was made as a round plate. An exit hatch was located in it, in front of the driver's seat. The radioman, who sat with his back to the driver, had a visor to the rear, and an exit hatch next to the motor. For the

The Maschinenfabrik Augsburg-Nürnberg delivered a vehicle that, after short testing, was accepted for series production (MG *Kampfwagen* (2 cm)(Vs.Kfz. 622).

commander, who was also the aiming gunner, a 2 cm KwK 30 and an MG 34 were mounted in the turret. For them, 180 and 1800 rounds of ammunition were carried. The cannon had a one-piece breech with a locking lever, and a barrel length of 1000 mm. The rate of fire was 280 shots per minute. The normal range went up to 1200 meters. The turret itself, moved by handwheels, had four visors, and a two-piece hatch in the roof.

The Type "a2" that followed it (chassis numbers 20 011 to 20 025) hand a handhold in the rear wall for better access to the ventilator drive. The generator had a suction socket and fresh air duct.

Type "3a" was built in two series (chassis numbers 20 026-20 050 and 20 051-21 000). The firewall between the motor and radioman was now unscrewable. A large bottom flap under the

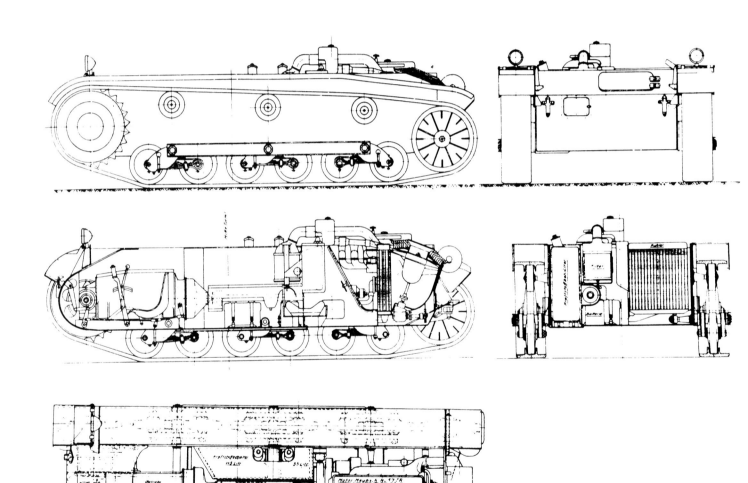

The first series vehicles of "Panzerkampfwagen II (2 cm) (Sd.Kfz. 121)" had the type designation "a1," "a2," and "a3."

motor served for removal of the fuel pump and oil filter. As of Type "a2," a welded leading wheel without a rubber covering was used. In vehicles 20 051 to 21 000, springs without extra leaves and with double bows were used. A radiator with 158 mm block depth was used. The fuel tanks, of which the first held 102 liters and the second held 68 liters, had a new filler with bayonet lock. In 1936 there appeared another version of "Panzer II," Type "2/LaS 100." This Type "b" (chassis numbers 21 001-21 100) had a changed bow, in order to hold the steering system with intermediate gears. In this arrangement, only the second planetary drive was replaced by a spur-gear system, which was now located on the outside of the armored hull. The external body was made of steel armor plate. The newly patterned drive wheel had, as before, a diameter of 755 mm. The road wheels were widened. New spring attachments with slit-head screws were used. The widened jack rollers had a smaller diameter. The track covering panels were lengthened in back and folded up. A new muffler—longer than

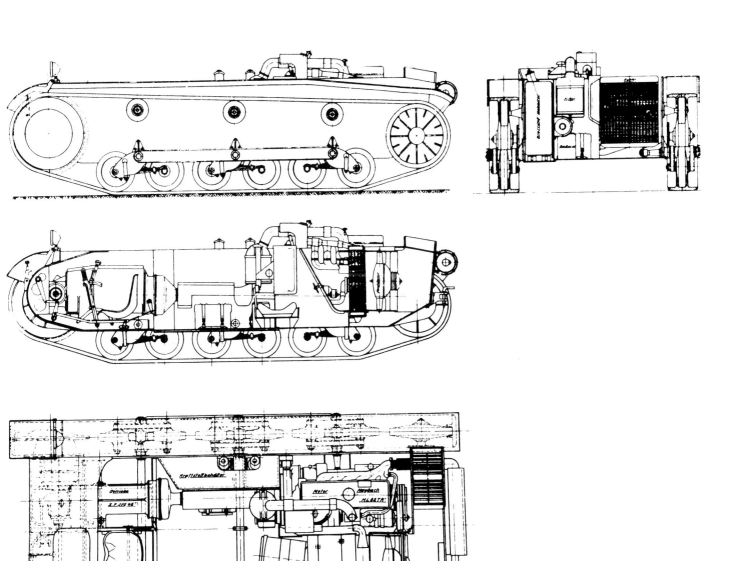

The Type "b" followed in 1936. Now the Maybach "HL 62" motor was used.

before, but smaller in diameter—was installed. The shift lever was shortened by 30 mm. Now the Maybach "HL 62 TR" motor was used, with 105 and 120 mm bore and stroke, giving a displacement of 6.2 liters, and producing 140 HP. The rear engine compartment was newly laid out.

In 1936 Type "c" (chassis numbers 21 101-2 000 and 2-001-23 000) appeared. As Type "3/LaS 100," this vehicle received a new hull for crank running gear. The five-wheel leaf-spring running gear was laid out so that each road wheel was mounted on a turning crank arm on the hull. Firmly attached to the crank was the leaf spring, the tapering end of which was braced against a turning roller on the hull. In suspending the road wheel, the crank pushed the leaf spring through under its jack roller, and thus increased the effective length of the spring. The track width, increased from 1780 to 1880 mm, required new jack roller axles, new drive wheels, and leading wheels of larger diameter.

Four different views of the Panzer II (Types a and b).

As in Type "b," the wide road wheels required new tracks.

Protection of the tracks was extended. Vertically mounted and relocated filler caps served the fuel tanks, which were thickened from 5 to 10 mm.

A new brake arrangement allowed an automatic setting of the track-brake support. Switching the headlight on for night driving was provided for. The total weight of numbers 21 001 to 27 000 had risen to 8.9 tons.

In numbers 22 020 to 22 044, molybdenum steel (substitute steel) was used. From 1937 to 1940 Types "A" (numbers 23 001-24 000), "B" (24 001-26 000), and "C" (26 001-27 000) were delivered.

The formerly curved bow plate was replaced by flat plates as of Type A. As of the fourth series, the front armor had been belatedly increased to 30 mm.

Henschel worked from 1937 to March 1938 on assembling these vehicles, with a monthly output of about 20 units.

The Altmärkische Kettenwerk GmbH (Alkett) was brought in as a daughter firm of the Rheinmetall-Borsig AG toward the end of 1937 for a monthly tank production (assembly) of, at first, thirty Panzer II tanks per month. The first finished tanks of these types were finished during 1938. These vehicles, of which some 300 in all were built at Alkett, were replaced by the Panzer III.

The front bottom reinforcement had been changed because of a new gearbox mount. This gearbox was the improved ZF "SSG 46" with synchromesh. An automatic setting of the track supports and support brakes was introduced. Two UKW radio receivers and one UKW sender were carried. The vehicle had appropriate interference elimination.

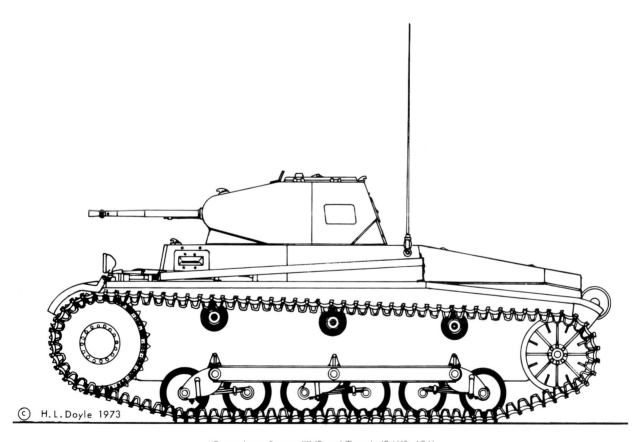

"Panzerkampfwagen II" (2 cm) Type b (Sd.Kfz. 121)

Front and rear views of "*Panzerkampfwagen* II" (2 cm) Type b (Sd.Kfz. 131).

This picture shows a tank of the 7th Armored Division supporting infantry units in France. These vehicles were used mainly during the campaigns in Poland and France.

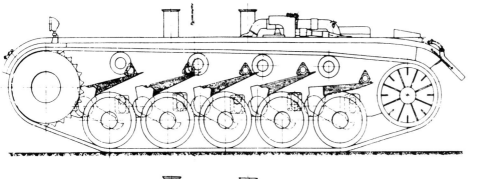
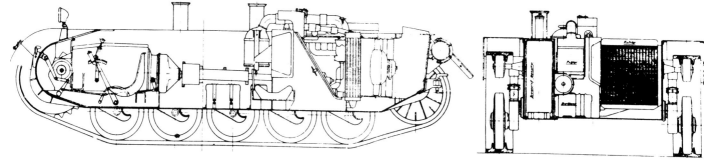

The "*Panzerkampfwagen* II" Type "c" appeared in 1937, and had the final 5-wheel running gear of the "Panzer II" series.

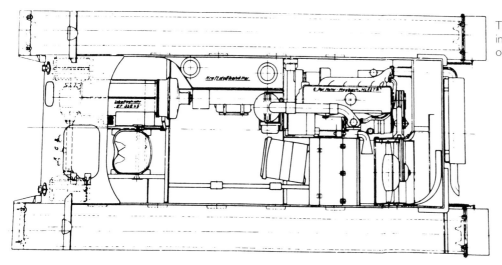

Typical of this vehicle were the round bow plate and the two piece turret hatch.

Panzerkampfwagen I (MG) (Sd./Kfz. 101), Type A.

Panzerkampfwagen I)MG) (Sd./Kfz. 101), Type B.

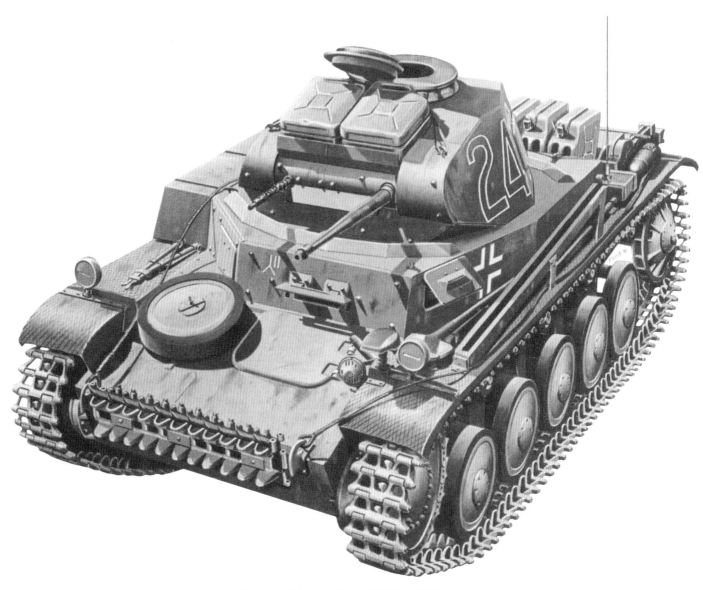

Panzerkampfwagen II (w cm) (Sd.Kfz. 121), Type C

Panzerkampfwagen II (2 cm) (Sd.Kfz. 121), Types D and E.

Panzerkampfwagen II (2 cm) (Sd. Kfz. 121), Type F.

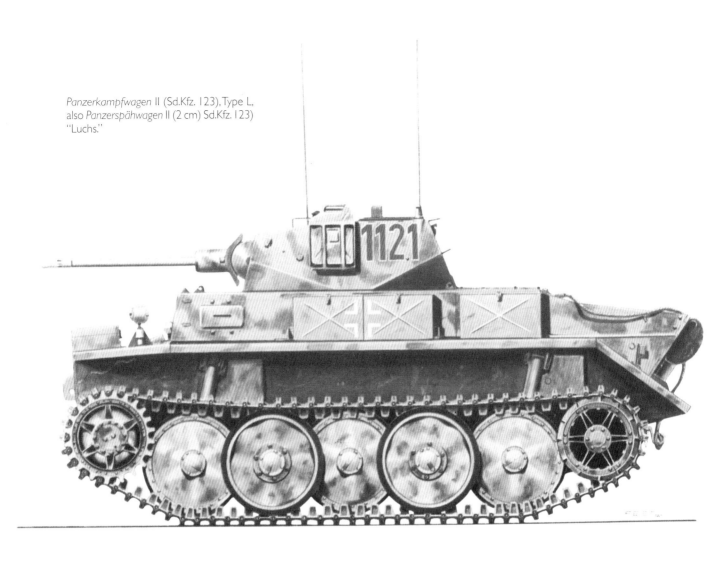

Panzerkampfwagen II (Sd.Kfz. 123), Type L, also *Panzerspähwagen* II (2 cm) Sd.Kfz. 123) "Luchs."

Leichte Feldhaubitze 18/2 on *Panzerkampfwagen* II chassis (Sf) (Sd.Kfz. 124), formerly "Wespe."

Details in the front end of the "Panzer II" show the gearbox next to the driver's seat, and the cardan shaft to the motor. The fuel tank is on the right side of the hull.

This picture shows the installed motor with the cooling system beside it. The two fuel tanks are easy to see.

The most important parts of the new "Panzer II" running gear, with road wheel, swing arm, leaf spring, and turning rocker.

The armored box body of "Panzerkampfwagen" II (2 cm), Type C:

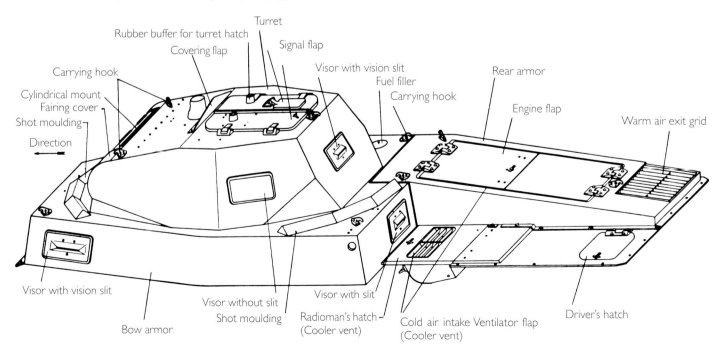

The armored body, seen from the rear. The picture shows the radioman's visor flap and the cutout for the antenna. The turret mounting is easy to see.

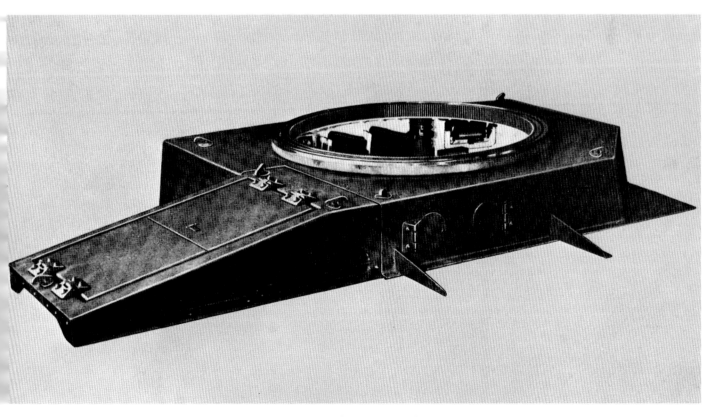
A look at the right side of the body, with the engine cover.

To protect the eyes against lead spray and splinters, protective glass was set behind the slits of the visors; in Types "A" and "B" it was 12 mm thick, from Type "C" on, 50 mm thick. As of Type "A," a flat commander's cupola was installed in the turret roof instead of the two-piece hatch. Its ring flange held eight periscopes, which allowed a 360-degree field of vision. The cupola itself was closed by a round flap. The "C" series went into production in March 1940.

In the meantime, the vehicles had proven themselves in Poland, where they were also represented in the greatest numbers, although the production had been cut back strongly. From July to December 1939 the following production figures for "Panzer II" vehicles applied:

July	August	September
9	7	5
October	November	December
8	2	—

In November 1938 the Weapons Office gave a contract to the Maschinenfabrik Augsburg-Nürnberg to build a 175-200 HP Diesel motor (Type HWA 1038 G) for use in "Panzer II." The production was planned to begin at the end of 1940.

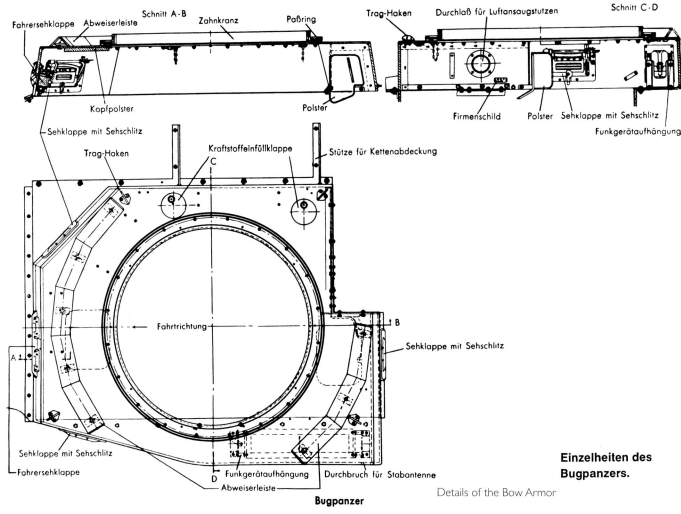

Einzelheiten des Bugpanzers.

Details of the Bow Armor

A memo of November 27, 1939, states that the original beginning of a new series of "Panzer II" vehicles was not possible because of the late arrival of the hull drawings. On November 1, 1940, five contracts for "Panzer II" were running; three of them went out in April and one in August 1940, but production did not begin until December 1940.

This new Type "F" (chassis numbers 28 001-29 400) had been given a completely new upper front armor and driver's area plate. The armor thicknesses had increased from the predecessors as follows:

	Earlier models	Type "F"
Turret front plate	14.5 + 20.0 mm	30 mm
Gun shield	14.5 + 14.5 mm	30 mm
Driver's front plate	14.5 + 20.0 mm	30 mm
Upper bow plate	14.5 + 14.5 mm	20 mm
Lower bow plate	14.5 + 20.0 mm	35 mm
Side armor	14.5 mm	20 mm

The total weight of the vehicle had thus increased to 9.5 tons. The straight lines of the upper plate over the driver's seat were noteworthy, which now had a dummy flap in front of the gearbox.

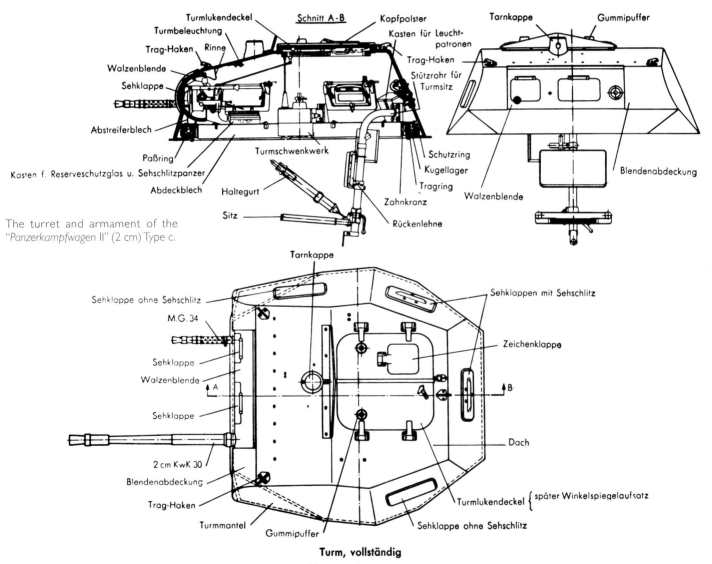

The turret and armament of the "Panzerkampfwagen" II (2 cm) Type c.

Turm, vollständig

The complete turret.

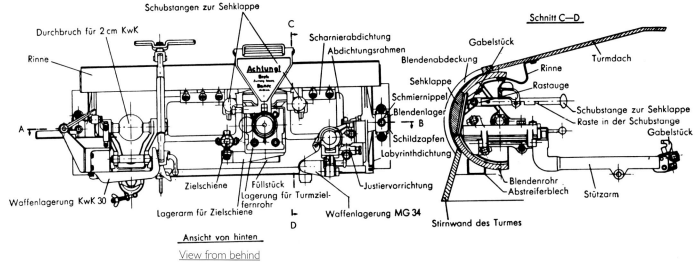

Cylindrical mount

Two views of the cylindrical mount with all details.

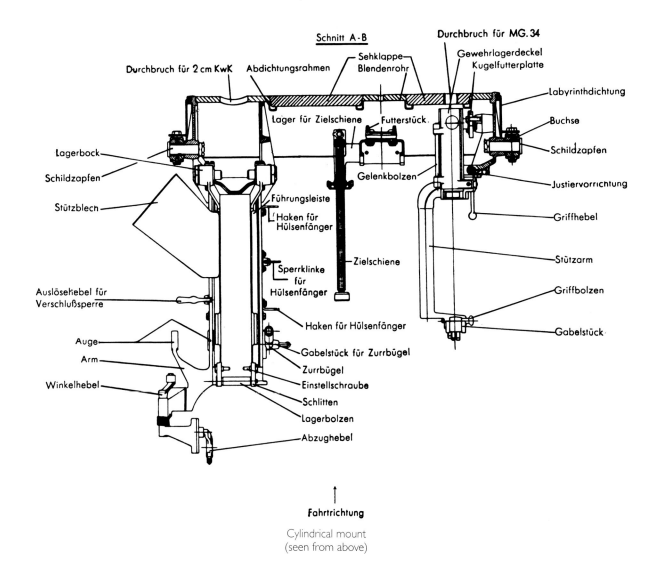

Cylindrical mount (seen from above)

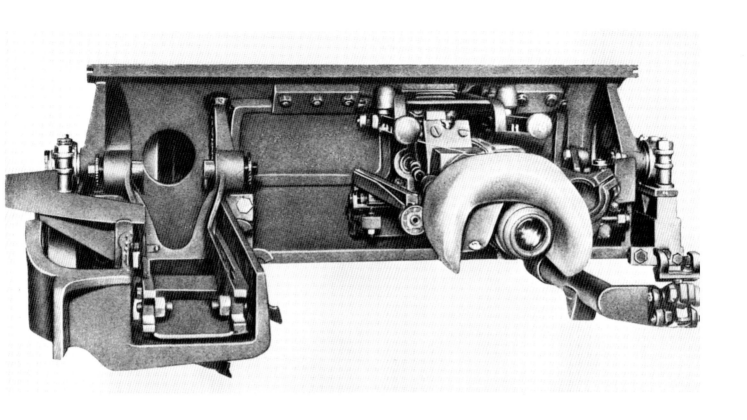

Inside view of the cylindrical mount, with 2 cm KwK mount at left and machine gun mount at right. In the middle are the aiming optics.

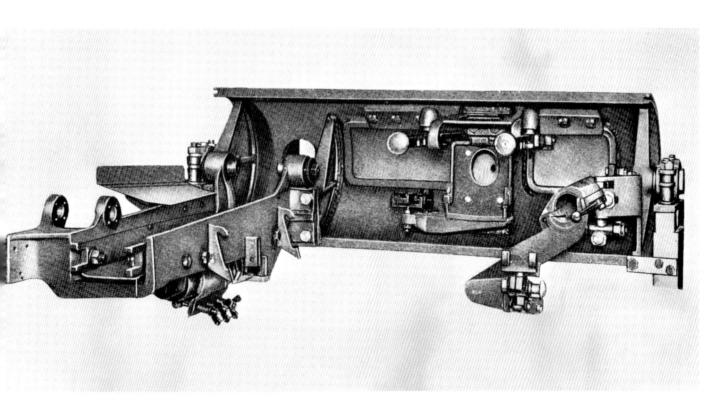

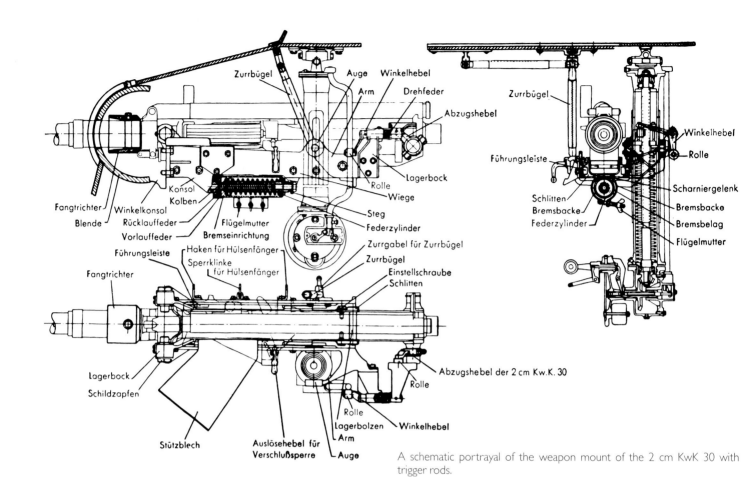

A schematic portrayal of the weapon mount of the 2 cm KwK 30 with trigger rods.

Type "c" of *Panzerkampfwagen II.* Side view. The tilted shot apparatus over the fender held the antenna.

Behind it, though, there was no opening, as the visor flap on the right side of the body had no use. Apparently, considerations for further use of this vehicle had been made but were never worked out later.

New leading wheels, now formed from conically shaped sheet-metal discs, also came to be used. The road wheels were made partly of aluminum. Armament and radio equipment remained unchanged, and the standard fog-cartridge launcher was attached to the back of the vehicle. The price of the tank (without armament) was 49,228 RM. In terms of armament, some of these tanks were fitted with the improved 2 cm KwK 38. This weapon had a turning-head breech and, like the KwK 30, a barrel length of 1000 mm. The number of riflings had been raised from two to eight. The effective ranges were: against ground targets to 1200 meters, against armor possibly to 1000 meters, and favorably against armor to 600 meters.

When the war began, the "Panzer II" formed the numerical backbone of the attacking armored divisions, and at the beginning

"Panzer II" (Type c), seen during the French campaign. The round bow of these tanks was fitted with extra armor plates as part of the generally carried-out armor strengthening.

The towing of broken down tanks was already practiced frequently in peacetime, and thus the troops had an excellent repair service on hand. The pictures show an 8-ton medium towing tractor (Sd.Kfz. 7) built by Krauss-Maffei (Type "KM m 9") with a low-loader trailer for tanks (Sd.Anh. 115) retrieving a "*Panzerkampfwagen II*."

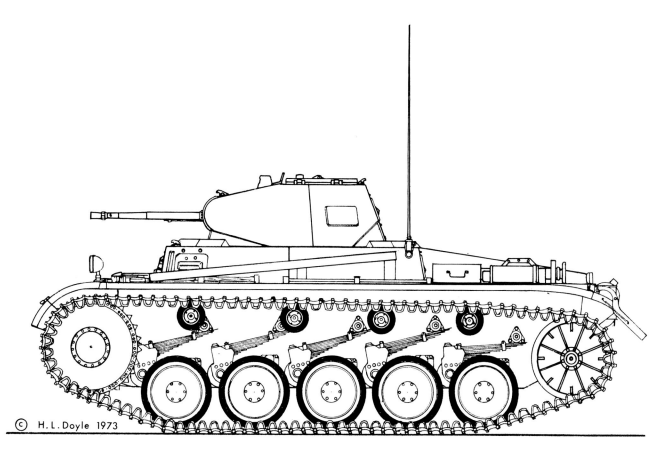

"*Pamzerkampfwagen* II" (2 cm) Type "c" (Sd.Kfz. 121).

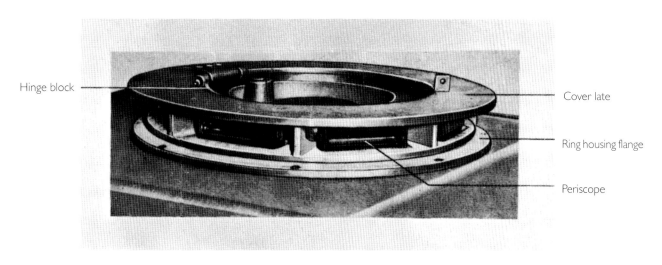

In Types A, B, and C of the Panzer II a flat commander's cupola was used.

Above: A look into the turret from below after installation of the flat commander's cupola. The weapon mount is easy to see. Below: A close look at the commander's cupola from inside shows the periscopes all around and the flap upholstery.

The low height of the cupola, seen close-up. The weapons have been removed.

A good look at "*Panzerkampfwagen* II" (2 cm) Type A (Sd.Kfz. 121).

Above: Additional armor was placed on the front of the "Panzer II" body and turret. Below: New vehicles of Type "B" are being accepted from the Maschinenfabrik Augsburg-Nürnberg.

The last production series of "*Panzerkampfwagen* II" was Type F. Its identifying marks are the dummy radioman's flap and the new track-tightening wheels, which are now conically shaped.

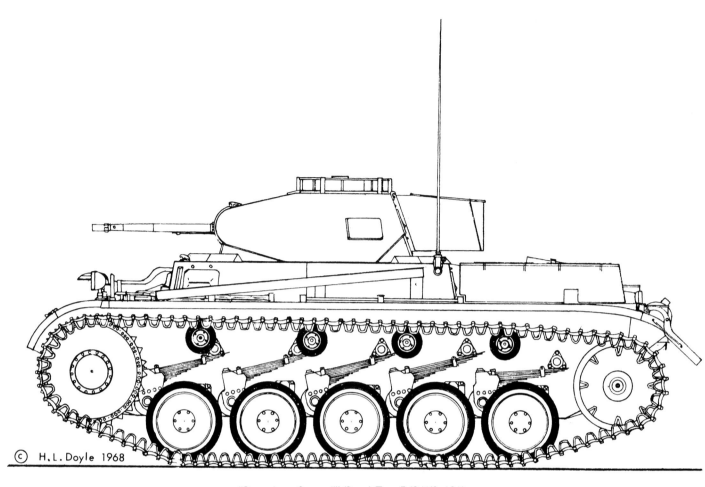

"*Panzerkampfwagen* II" (2 cm), Type F (Sd.Kfz. 121).

Below: These vehicles were used by the troops for years, and formed the backbone of the attacking armored divisions in North Africa, Yugoslavia, and Russia. They were fully overwhelmed after the appearance of the Russian "T 34."

of the French campaign in 1940 there were 955 "Panzer II" on hand. On July 1, 1941, there were still 1067 "Panzer II" on hand in the Army; this number had sunk to 860 by April 1, 1942. The "Panzer II" tanks used in North Africa had a different cooling arrangement and additional dust security. After such revisions, the designation was "*Panzerkampfwagen* II (Tp)." In 1938, two further series of "Panzer II" vehicles were already planned, which had new running gear. They used four large road wheels, fully rubber-tired and sprung on torsion bars, which made jack rollers unnecessary. The basic division of space was also changed, giving

A tank on duty in Russia. In the air is a communication airplane of the Fieseler "Storch" type.

the radioman a seat next to the driver in the front of the vehicle. He had a visor in the front of the body, and a second hatch in the upper bow plate. The air-suction gearbox of Maybach Variorex "VG 102128 H" type, with seven forward and three reverse speeds, was attached directly to the motor, and linked by a driveshaft with the intermediate gears at the front. The drive wheels were in the front.

The Daimler-Benz AG in Berlin-Marienfelde was responsible for this series, and delivered Type "D" (chassis numbers 27 001-27 800) with the type designation "8/LaS 138." The vehicles were mainly sent to light divisions as "*Schnellkampfwagen.*" The immediately following Type "E" (chassis numbers 27 801-28 000) was almost identical, with the exception of the tracks, which were now greased.

Above: The "D" type of "*Panzerkampfwagen* II" had four large road wheels on each side, suspended on torsion bars. The radioman's seat was now next to the driver's.

Left: "Panzer II Type D" being loaded on 6-wheel Büssing-NAG trucks via Special Trailer 115. These tanks belonged to the equipment of the "Fast Divisions."

Two hundred fifty of these tanks were produced. The Maschinenfabrik Augsburg-Nürnberg produced 68 of them. They had a total weight of ten tons, and a top speed of 55 km/h. The unchanged "Panzer II" turret was used.

The main supplier of hulls, upper body parts, and turrets for the "Panzer II" building program was again the Deutsche Edelstahlwerke AG in Hannover. Their deliveries between 1936 and 1942 ran as follows:

	Hulls	**Bodies**	**Turrets**
1936	117	147	84
1937	215	309	194
1938	308	346	432
1939	—	85	2
1940	42	118	118
1941	132	92	92
1942	148	172	54

The role of the "Panzer II" in Operation "Sealion" should also be mentioned. It is known that Panzer Units A and B were made up at Putlos in September and October 1940 of volunteers from Panzer Regiment 2, who were trained for the invasion of England. For that, the "Panzer II" assigned to the units were turned into amphibian tanks by special equipment. WaPrüf 6 had ordered floats from the firms of Alkett in Berlin, Bachmann in Ribnitz, and Gebr. Sachsenberg in Roslau, to give the "Panzer II" a water speed of 10 km/h, and a sea capability even in seaways 3 or 4. Fifty-two of these floats were ordered, and were to be fastened to the jack rollers of the tanks. The containers were divided into three compartments and filled with celluloid tubes. Through these containers the drive went to the two rear screw propellers, which were driven by the leading wheels via extension sleeves, cardan gears, and shafts. Between turret and hull a sealing tube had been installed. In the water, the tank was immersed about up to the track covers. The tanks remained fully ready for action while floating.

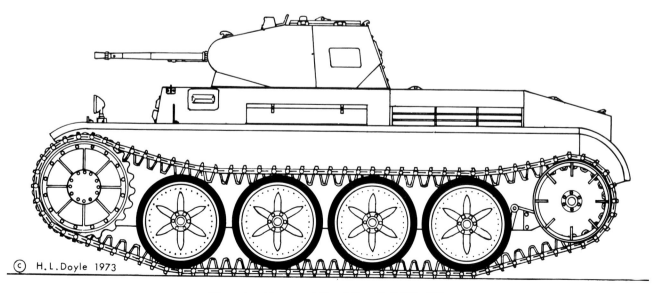

"Panzerkampfwwagen II (2 cm), Type D (Sd.Kfz. 121).

"Panzer II" in action. By the attachment of side floats, the tanks became sea-capable. The weapons remained ready to fire in action.

The floating "Panzer II" originally prepared for Operation "Sealion" were used later in Russia. The picture shows a vehicle with the extending sleeve for the drive, as well as the brackets for attaching the floats.

The floats were made by Gebr. Sachsenberg in Roslau. The Kassbührer firm in Ulm also took part in this project.

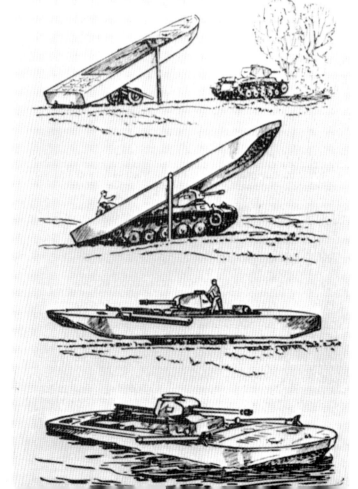

A further attempt to make the "Panzer II" amphibious is shown by this suggestion from Gebr. Sachsenberg in Roslau.

On the basis of a conference with Hitler on July 7, 1941, it was considered purposeful that all tanks of future new production were to be reinforced by a pre-armor—separate from the armor—in order to counteract the penetrating power of the new hollow-charge shells. The expected increase in weight and loss of speed were to be accepted in the bargain, in Hitler's opinion (GFM Keitel to OKH). For the "Panzer II" vehicles still fitted with a round bow plate, additional 20 mm plates were attached to the front. The added bow plates were made in two pieces, the upper ones 14.5 mm thick, the lower ones 20 mm. Thus, the tanks came to look similar to Type "F." Additional curved 20 mm plates also appeared on the turret shield. In this respect it is interesting to note an excerpt from a meeting of the Panzer branch on July 17, 1941. New instructions from the *Führer* requested an increase of the Panzer troops to 36 armored divisions. A representative of the General Army Office determined that to set up these divisions, among other things, 4608 "Panzer II" were necessary. It is surprising that they still stuck with the production of "Panzer II," although the experiences of the previous campaigns already showed clearly that these tanks were fit only in exceptional cases to fight against enemy tanks. The "Tank Program 41," which covered the time to 1949 in its future plans, requested as primary equipment, in view of the need for yearly replacement and spare parts for the peacetime army, among other things, 18,946 *Panzerkampfwagen* II. Adding replacements, this amounted to a need for 21,860 tanks. It had already been considered that the present "Panzer II" models were to be replaced by the "VK 903" still to be mentioned.

These plans allowed the further development of "Panzer II," and thus the Types "G1," "G2," "G4" (chassis numbers 150 001-150 075), and "J" tanks (chassis numbers 150 101-150 130) were to follow into 1942. Only prototypes of them were built.

In 1941 the AHA/AgK (Inspection 6) advocated a tank of the ten-ton class with "increased speed and better armor." A development chassis of this Panzer II (Types "H" and "M") was delivered by the MAN firm of Nürnberg on September 1, 1941. Fitted with a Maybach "HL 66 P" six-cylinder gasoline engine of about 200 HP, the 10.5-ton vehicle was to have a top speed of 65 km/h. The armor was to be 30 mm thick in front and 20 mm on the sides and rear. The roof was protected with 10 mm plates, the bottom with 5 mm plates. The track was 2080 mm. The three-man crew had a 2 cm KwK 38 and an MG 34 to use. Production was scheduled to begin in mid-1942, but the concept was passed over due to the events of the war at that time.

The monthly production quota of "Panzer II" tanks was set at 45 units as of 1942. Shortages of workmen, especially at FAMO (which also produced these vehicles as of 1940 at the Vereinigte Maschinenwerke in Warsaw), resulted in considerably lower totals. In 1942 Daimler-Benz produced 18 tanks, MAN made 40, and FAMO built 382 "Panzer II" chassis. Until production was halted in 1944, a total of 623 tanks of this type were built.

On June 18, 1938, the Weapons Office had given the Maschinenfabrik Augsburg-Nürnberg AG (chassis) and Daimler-Benz AG (bodies and turrets) contracts for the further development of *Panzerkampfwagen* II with "emphasis on high speed." This vehicle was designated "*Panzerkampfwagen* II n.A. (VK 901)." The first chassis was finished at the end of 1939, and fitted with the six-cylinder, 145 HP Maybach "HL 45" motor. With front armor of 30 mm and 14.5 mm on the sides, the total weight was 9.2 tons. The top speed was 50 km/h. 60 km/h was striven for, but for that a 200 HP motor was required, which was later available as the "HL 66 P." The three-man crew had a 2 cm KwK 38 and an MG 34 in stabilized mounts in the turret to use. The 0 series of 75 tanks began in October 1940 and was delivered.

An improved version of this "fast runner" as "VK 903" was to be fitted with the Maybach "HL 66" motor and "SSG 48" gearbox. A top speed of 60 km/h was striven for. An improved version of the Panzer 38 (t) transmission and steering was planned for the "VK 903."

Thirty "VK 903" vehicles, for which contracts were given to Daimler-Benz, Rheinmetall-Vorsig, and Skoda on June 1, 1942, were to be rebuilt as armored observation vehicles for motorized artillery and armored regiments. One of these "VK 903," planned as standard equipment with the Cupola 1303 b, was finished by September 1942 as a test vehicle, and also fitted with a range finder, radar locator, and observation and radio equipment. The Army's focal point program (Panzer Program 41/Army Flak Program) of May 30, 1941, called for, among other things, the use of "*Panzerkampfwagen* II n.A. (VK 903)" in greater numbers. The running gear of this vehicle had been designed by Dr. Ing. h.c. F. Porsche (Porsche type 168). With all the planned variants, and in view of the need to catch up for the Replacement Army and supplying, there was a total need of 21,860 of these vehicles in 1941. The program foresaw the following variations: *Panzerkampfwagen* for combat reconnaissance (3500); for reconnaissance (armored scout cars), (10,950); light tank destroyer

A variant of *Panzerkampfwagen* II designed for high speed was the "VK 901," built by MAN. It was designated "*Panzerkampfwagen* II neue Ausführung."

(Pz. Sfl. 5 cm) for antitank units of armored divisions (2738); for fire control and observation in artillery and motorized divisions (2003); and for heavy infantry guns (Pz Sfl) (481 units).

Because of insufficient performance, this program was not developed further.

On December 22, 1939, another contract went to MAN and Daimler-Benz for the further development of Panzer II with "focal point of heaviest armor." The type designation of this vehicle was "*Panzerkampfwagen* II n.A. verstärkt (VK 1601)." A 0 series of 30 units was to begin, with delivery scheduled to begin in December 1940. The first chassis was running on June 18, 1940, and the first turret was finished on June 19, 1940. An order for the delivery of a first series of 100 units was withdrawn.

These vehicles were also fitted with the Maybach "HL 45" powerplant, which gave the vehicle, weighing up to 17 tons, a top speed of 31 km/h. The track measured 2350 mm, and the armor was 80 mm thick in front and 50 mm on the sides. The three-man crew had a 2 cm KwK 38 and an MG 34 in stabilized mounts in the turret.

Large numbers of this tank were also foreseen in "Panzer Program 41." There was to be a heavily armored tank for combat reconnaissance (339 units). This vehicle was later developed further as VK 1602 (Leopard). There was also to be a flamethrowing tank (259 units).

A notice of April 1941 states the following on this subject: "...In a conference at Daimler-Benz on April 19, 1941,, re the *Panzerkampfwagen* II n.A. Type (F) and Renault B 2 (f) tanks as heavy flamethrowing vehicles, Dircetor Wunderlich expressed that he had always been able to grant the wishes of the WaA for suddenly appearing constructive investigations, but it was no longer possible when such contracts were also given by other parties than WaPrüf 6. He noted that about four weeks ago a contract had been issued to Mercedes-Benz via Director Hacker to investigate the mounting of the 4.7 cm Pak (t) in the *Panzerkampfwagen* II turret...." Such an action did not take place.

Experience with the "VK 901/903" and "VK 1601" vehicles resulted in the design of a "*Panzerkampfwagen* II n.A. (VK 1301)." This contract from the Weapons Office had defined a version looking and measuring just the same as the "VK 901." A soft-steel prototype had been finished by the end of April. The fighting weight was 12.9 tons. After minor improvements, the vehicle went into series production as "VK 1303." The urgent need came mainly from the failure of wheeled armored vehicles for reconnaissance in Russia. It had been recognized early on that these vehicles could be used well in western Europe, but in the east they were only of limited use. Thus, on September 15, 1939, an order for the creation of an armored full-track reconnaissance vehicle with medium-range radio and radiotelephone equipment was issued by the AHA/AgK/In 6 to the Weapons Office.

Eight hundred of these vehicles were then ordered without an introduction contract. The development firms were the Maschinenfabrik Augsburg-Nürnberg for the chassis and Daimler-Benz for the body and turret.

The weight of the production version was 11.8 tons. Armor steel plates were used for the hull, and the welded body followed the usual building requirements. The stepped front of the vehicle was protected by 30 and 20 mm plates. The side plates were 20 mm thick; the same thickness was used at the rear, and the bottom of the hull measured 10 mm. The front of the body was screwed onto the hull using angle irons. Insulation between the hull and body prevented dust and water from entering. The body itself was closed at the top by 12 mm plates. The rear part of the body, including the engine hoods and air intakes, was also screwed on, and could be removed completely for major repairs. A fairly large turret with 360-degree traversing field was located centrally on the front part of the body. A commander's cupola was not included. The turret itself was armored with 30 mm plate in front and 20 mm on the sides and back. A rectangular entry hatch was in the back of the turret. The turret roof, which sloped down somewhat toward the front, was 12 mm thick. In the rear part of it there was a round exit hatch, plus a smaller opening for signal connections. The primary weapon was a 2 cm KwK 38, plus a coaxial MG 34. Aiming the weapons was done with handwheels. The elevation field extended from -9 to +18 degrees. Thirty-three magazines with a total of 330 rounds, plus 2250 rounds of MG ammunition, were carried in the tank. The firing height was 1690 mm, and there was a *Turmzielfernrohr* 6 for targeting. The crew consisted of four men.

The rear-mounted motor of Maybach "HL 66 P" type produced 180 HP at 3200 rpm. This water-cooled 6-cylinder engine had a displacement of 6754 cc. A governor prevented over-revving the engine. A cooling water exchanger allowed the adding of pre-warmed coolant from other vehicles during low outside temperatures. The engine power was transmitted by a dry two-plate clutch to the ZF Aphon "SSG 48" gearbox. A reduction gear was mounted between the clutch and gearbox. Six forward speeds and one reverse were available.

The "*Panzerkampfwagen* II new type, strengthened" (VK 1601), made by MAN, emphasized heavy armament. This vehicle was intended for combat reconnaissance.

Power from the gearbox was transmitted to clutch steering, which affected the front-mounted intermediate gears. Outer band brakes were used.

The radio devices were an 80-watt sender, a medium-range receiver, a radiotelephone "A" and an on-board speaker.

This vehicle was designated "*Panzerkampfwagen* II Ausf, L (Sd.Kfz. 123)." Its exclusive use by reconnaissance units changed the official designation to "*Panzerspähwagen* II (2 cm)(Sd.Kfz. 123), Luchs" (Lynx). Chassis numbers began with 200 100. After 100 vehicles had been built with the 2 cm KwK, it was planned to switch to the 5 cm KwK L/60 as of vehicle number 200 201 (April 1943). Thirty-one of these vehicles equipped with an open turret were built. In all, MAN produced 113, and Henschel built 18 of these vehicles. Production was halted on May 12, 1943.

The insufficient armor and armament of the "Lucks" reconnaissance vehicle moved the Weapons Office to set new building requirements for a heavily armored tank for combat

The "VK 1301" was the forerunner of a full-track reconnaissance vehicle. Only a soft-steel version of this vehicle was built.

reconnaissance. By going back to the already noted "VK 1601," the MIAG firm in Braunschweig received a contract in 1941 to develop the "*Gefechtsaufklärer* VK 1602." MIAG was responsible for the chassis, while Daimler-Benz provided plans for the turret and body. The project was called "Leopard." It is noteworthy that the fighting weight of this vehicle was set at some 26 tons. Armor measuring 50 to 80 mm was planned for the turret, and 20 to 60 mm plates for the hull and body. A 12-cylinder Maybach "HL 157" gasoline engine producing 550 HP was the planned powerplant, and was to give the vehicle a top speed of 60 km/h. The outer dimensions of the vehicle were 6450 x 3270 x 2800 mm. The width of the ungreased tracks was 650 mm. With 3475 mm of the tracks on the ground, the ground pressure could be reduced to 0.49 kg/cm^2.

The drawings for the armored parts of the chassis were finished on June 30, 1942, and those for the main components of the chassis on September 1, 1942. The chassis assembly drawings were supposed to be finished on November 1, 1942. At this time, though, the project was already outmoded, and no prototype was built.

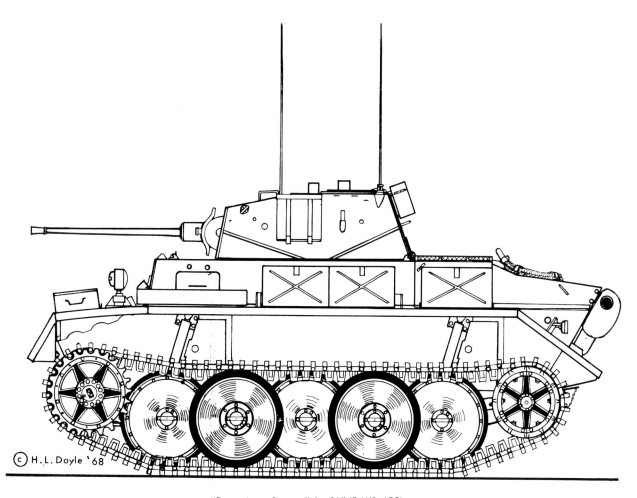

"*Panzerkampfwagen* II Ausf. L" (Sd.Kfz. 123).

The "*Panzerkampfwagen* II Ausf. L" (Sd.Kfz. 123) with the evocative name of "Luchs" (Lynx).

These vehicles were used by some reconnaissance units, mainly in Russia.

Details of the drive wheel and box-type running gear of the "Luchs."

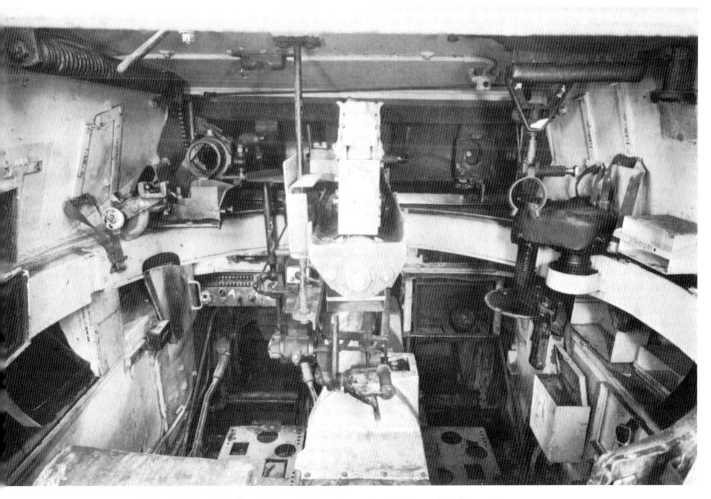

An inside view of the "Luchs" turret shows the central location of the 2 cm primary weapon.

The vehicle's turret, though (a wooden prototype of which was finished at the end of May 1942), later went into production in a slightly changed form. It was used on the 8-wheel "ARK" vehicle made by Büssing-NAG (Sd.Kfz. 234/2). The welded design used 60 mm plates in front and 40 mm for the rest of the turret. The roof was protected with 30 mm plates. The turret sloped downward to the back to allow the installation of radio sets. The 5 cm KwK 39/1 L/60 made by Rheinmetall was mounted in a "pig's head" shield. There was also an MG 42 in the turret. The elevation range went from -10 to +20 degrees. Two crew members (the commander/gun loader and the aiming gunner) were in the turret; the driver and radioman sat in the hull. The weapons were aimed by using handwheels. The turret crew had two exit hatches in the turret roof. A "*Turmzielfernrohr* 9" aiming scope was used.

As before, the "VK 1601" was the basis of plans in March 1942 for a combat reconnaissance vehicle (small type) with heavy armor and high speed. A 5 cm KwK L/60 with at least 60 rounds was planned. The armor strengths were 80 mm front, and 50 mm sides and rear. The ground clearance was to be at least 50 cm. Especially good vision and observation facilities were required for the four crewmen. The radio equipment was like that of the armored scout car. This development was carried on by Porsche and Skoda,

The last vehicles of the "Luchs" series were armed with the 5 cm KwK L/60. The turret was open at the top to save weight.

but the project, like that of the "VK 1602" combat reconnaissance car, was given to MIAG in Braunschweig. Speer reported in October 1942 that the troops, in reference to the "Leopard," preferred a more lightly armored, and thus faster vehicle. The "small building type" of the vehicle was thus preferred. Hitler noted that the armor of a reconnaissance vehicle had to be judged from different points of view. But since the "Leopard" in its heavier version had come too near the "Panther" in terms of driving characteristics, the sense of producing both vehicles was doubted. Besides, the 5 cm gun was insufficient for such a vehicle. Hitler said he agreed to put the "Leopard" into production only in its lighter version (VK 1601), under the assumption that the "Panther" should be used as a reconnaissance vehicle for some armored units. In January 1943 the production of the "Leopard" was canceled before it had gone into that phase. Neither in armor nor in armament (including as a reconnaissance vehicle) did it correspond to the conditions that were expected in 1944.

The idea of creating a well-armored, fast combat reconnaissance vehicle was followed up all the same. This was the "VK 2801" project, for which Daimler-Benz in Berlin-Marienfelde was responsible. It was a so-called multipurpose tank, a light, fast vehicle, which was to have a rear-mounted, air-cooled, 525 HP Diesel engine. The vehicle was to have several bodies. As long as the motor, still to be developed, did not exist, the 450 HP "DB 819" Diesel was suggested. Prof. Nallinger, though, suggested the immediate use of the larger "MB 507" motor, since the suggested 17-liter "MB 819" could be used as a tank engine with a steady performance of only 400 HP. Since the vehicles were to get heavier and heavier, and the performance of the cooling system and all other components usually were not added in, he believed that some 600 HP should be available in the end.

But the Weapons Office insisted on having the 12-cylinder "MB 507" developed into an eight-cylinder powerplant, which would fit better into the limited space. The fact that a new crankcase and crankshaft would have to be created made the hesitance of Daimler-Benz understandable. Supercharging was considered.

In October 1943 it was determined that, in the meantime, the total weight of this vehicle had risen to some 33 tons. The specialist responsible for the new project under the Inspector-General of the Panzer Troops originally wanted the 700 HP Maybach "HL 230" gasoline engine to be the basis of the new design, since the Daimler-Benz "MB 507" would be some 350 mm longer, and thus would cause a further increase to the original calculated weight of 28 tons. Another standpoint for the preference of the "HL 230" would be that the change from aluminum to cast iron in the Maybach motor was already made, as was the replacement of roller bearings with journal bearings.

On November 8, 1943, it was reported that, under the conditions, it had to be expected that work on the multipurpose tank would be stopped. At that time, the responsible authorities were having conferences about the future tank program, since the Reich Minister for Armament and War Production had declared on October 9, 1943, that, for concentration of development in the area of armament and war production, it should all hang together.

It was learned from the OKH that recently the development of a fast, light vehicle with strong armament was planned, which should, in the interest of a short-term series production, be built mainly of components from the *Panzerkampfwagen* III developed by Daimler-Benz. Daimler-Benz had already made preliminary investigations on the subject of such a vehicle, and had discussed them with the OKH.

These discussions had led on November 22, 1943, to the decision that the project should be shelved in favor of a different solution, in view of the relatively high total weight.

This new design should be made on the basis of the Panzer II, and FAMO should be responsible for it.

Daimler-Benz should again take up the work on the multipurpose tank, the further development of which was assured for the time being. At first the individual components to be used in the multipurpose tank, along with the "HL 230" gasoline engine, should be selected; then the same investigations should be made on the basis of the "MB 507" Diesel engine.

On May 8, 1944, a message stated that the development of the multipurpose tank (VK 2801) had meanwhile been postponed temporarily by the Army Weapons Office. The project was thus finished once and for all.

A so-called "*Panzergerät* 13" was also mentioned in reference to the "Panzer II" series, but no further references to it have been found.

"*Gefechtsaufklärer* VK 1602" (Leopard) is seen with a test command in Thuringia in 1942 on the occasion of its 2000th kilometer of driving tests.

Before having a closer look at the variations of the "Panzer II," it should be mentioned briefly that the "Panzer II," like the "Panzer I" before it, had been loaded onto trucks and low-loader trailers for road transport by the fast divisions. The Büssing-NAG "654" and "900" truck types and the Faun "L 900/D 567" were mainly used. So was the Low Loader Trailer for Armored Vehicles (Sd.Anh. 115), which had been made in two versions for eight- or ten-ton loads.

A series of flamethrowing tanks for use in special armored units had been contracted for by the WaPrüf 6 on January 21, 1939. The development contracts went to the MAN firm in Nürnberg for the chassis, and to Wegmann & Co. in Kassel for the body and turret. By using unchanged hulls of the "8/LaS 138" type, the body had been fitted with a smaller machine-gun turret after the canon turret had been removed. A small turret had been mounted on each front fender, which was equipped with a flamethrower. The traversing range was 180 degrees. The burning oil supply, carried in armored tanks on the outside, was sufficient for About 80 2- to 3-second bursts of flame. The range was some 35 meters. Two-man crews were planned. A total weight of 11 tons was reached. An "A and a "B" type were built. The production of this "*Panzerkampfwagen* II (F) (Sd.Kfz. 122)" began as of January 1940. The test series of 87 units plus three spares was to run until October 1940. On June 19, 1940, there were sixteen on hand; the last nine were delivered in January 1942.

The tactical value of these vehicles left much to wish for, and they were returned on December 20, 1941, when the Army Weapons Office issued a contract to create a tank destroyer (Sfl). The chassis of the "LaS 138" were to make captured Russian 7.62 cm Pak and leFK guns mobile. This wartime solution was given to Alkett without an introduction request. The 7.62 cm Pak 36 (r) had a barrel length of 2179 mm (L/54.8) and a range of 7.6 km. The maximum initial velocity was 920 m/sec. The body of the vehicle, its open-topped structure protected by 14.5 mm plates, was rather high, measuring 2600 mm. The firing height was 2.2 meters. A traversing range of 50 degrees was available. The elevation went from -5 to +16 degrees. Thirty rounds of ammunition and a four-man crew completed the equipment of 11.5-ton vehicle.

By May 12, 1942, 150 of these "Pz. Sfl. I für 7.62 cm Pak 36 (r)" had been turned out and given the evocative name of "Marder" (Marten). A contract for another 60 sets of armored bodies was issued, but the further production depended on the delivery of repaired Panzer II (F) chassis. On one series of these vehicles the Russian 7.62cm Field Cannon 296 was also installed, without a muzzle brake. It was also possible to mount the German 7.5 cm Pak 40/2 on it.

Hitler made known on May 13, 1942, that the production of Panzer II amounted at this point to an average of 50 per month. Speer doubted whether the vehicles would ever have any fighting value for the troops. A suggestion to use the chassis as self-propelled mounts for the 7.5 cm Pak 40 was put forth. Thereupon the Reich Minister for Armament and Ammunition gave an order (No. 6772/42 g) to the Weapons Office on May 18, 1942, to create another tank destroyer on self-propelled mount. This development, promoted by Hitler, was turned over to Rheinmetall-Borsig for the gun, Alkett for the body, and MAN for the chassis. By using all types of "LaS 100" chassis, the installation of the 5 cm Pak 38 on a few chassis was tried at first. For series production, though, the 7.5 cm Pak 40 was used, now modified as Pak 42, with the "Panzer II" chassis, as "Sfl II für 7.5 cm Pak 40/2" (Sd.Kfz. 131). The gun, with its upper mount, was set on and screwed fast to the top of the upper armored body by using a newly made lower mount (a plate with a toothed ring).

The gun crew (aimer and loader) were protected at the front and sides by a firm armored structure, and secured by a swinging gun shield. The fighting compartment was open on top and in back. The body armor was safe from SmK fire.

The "*Panzergerät* 13," of which no other information exists.*

Opposite: The "*Panzerkampfwagen* II (F) (Sd.Kfz. 122)," rebuilding of which was done by the Wegmann firm. This picture shows the flamethrower turrets on the front fenders.

* During printing, further research showed that the vehicle in the picture is the VK 601 made by Krauss-Maffei (outer disc wheels removed).

While on the move, the gun was lashed down to the cradle in back and the support of the barrel in front, to protect the toothed ring.

The gun's ammunition was carried on the rear of the vehicle in a three-section armored ammunition box: in the left side 24, in the middle 7, and in the right 6 shells were carried.

The traversing field was about +32 to -25 degrees, the elevation from +10 to -8 degrees. The firing height was 1940 mm. There were also an MG 34 and an MP 38 for the three-man crew. The radio equipment included a "d" radiotelephone and a speaker. For protection from dust and rain, the fighting compartment could be covered by a canvas. The crew consisted of the aiming gunner (also gun leader), the loader, and the driver, who was also the radioman.

The total weight of the vehicle was 10.8 tons. On June 15, 1942, the first vehicles were delivered, to be followed by 1216 more. These vehicles also bore the evocative name of "Marder II."

The Armored Self-propelled Mount I for 7.62 cm Pak 36 (r) (Sd.Kfz. 132) also had the ecovative name of "Marder II."

Despite the open-topped fighting compartment, these guns were a valuable help to the antitank defense in Russia.

At the beginning of June 1942, Field Marshal Keitel was commissioned to determine whether the whole production of "Panzer II" should be switched to self-propelled mounts. Hitler, however, released only half of the production for this use.

Toward the end of June 1942, 75% of the Panzer II production had been released for self-propelled mounts. If the Army needed even more of these vehicles then, according to Hitler, the whole production should be switched to them.

In order to increase assault-gun production, the elimination or reduction of Panzer II self-propelled mount construction was considered in September 1942. Hitler insisted that these vehicles were not as popular among the troops as the assault guns. The vehicles proved to be too weak. In November 1942 tests were made to strengthen these makeshift solutions, which led to no success, mainly because of the limited power of their motors.

Of the already noted *Panzerkampfwagen* II n.A. (VK 901), six more of which were made in January 1942, two chassis, according to an order from the Weapons Office on July 5, 1940 (Az. 73a/p AgK/In 6 (VIIIa) No. 1684.40 g), had been rebuilt into light tank destroyers.

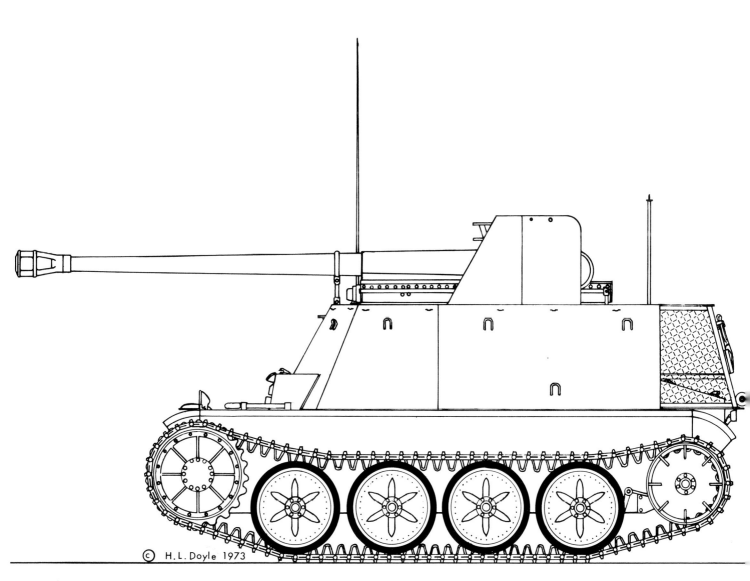

Pz Sfl for 7.62 Pak 36 (r) (Sd.Kfz. 132).

The open-topped body was typical of these vehicles. But for several years, along with other versions, they formed the backbone of the German Army's antitank defense.

The crew gets aboard the vehicle during combat training. The high bodies of these vehicles limited their combat value.

The two test vehicles, which reached the front in January 1942, were designated "5-cm-*Geschütz* auf *Panzerkampfwagen* II *Sonderfahrgestell* 901" (Pz.Sfl. 1c). The Rheinmetall firm in Düsseldorf was responsible for the gun and body, while the chassis modifications were done by MAN in Nürnberg. With a total weight of 10.5 tons, front armor of 30 mm, and side armor of 20 mm, the four-man vehicle was hopelessly inferior to the Russian T 34 tank in combat. The tests were not continued.

In July 1942 it was requested that the mounting of the 8.8 cm Pak 41 on the Panzer II chassis be tried. Should this combination not be possible, then at least the 7.5 cm L/71 cannon should be mounted. Under any conditions, the new motor should be used for these vehicles. These projects were not carried out.

The urgent need to give the armored units the necessary artillery support led to the introduction of so-called "gun wagons." To save time, as many production components as possible should be used. During the final solution of special development for the armored artillery, the interim types should be created as soon as possible. For the light 10.5 cm howitzer the "Panzer II" chassis was chosen, and all the production of that vehicle was devoted to that use. The original armor of the chassis was kept, and that of the body was made SmK-safe. The connection pieces between the gun and the vehicle were to be done simply and quickly. The machining and use of fitting was to be limited to the most necessary.

The Alkett firm in Borsigwalde had received a contract to create a makeshift mount for the light howitzer by using available gun and vehicle parts. Under these technical conditions it was not possible to give the makeshift mounts all-round fire, nor the ability to dismount the gun. The high power-to-weight ratio required for self-propelled artillery mounts was also not to be attained, since available powerplants had to be used. In cooperation with MAN, it was possible to mount the motor right behind the driver in a short time, thus freeing the rear part of the vehicle for mounting the gun. In addition, 32 rounds of ammunition could be carried.

The communication equipment of the makeshift mounts was already set up so that the organizational command experience could be used for the final solutions. The armored artillery observation wagon (Panzer III makeshift solution) had the same final communication equipment.

The "Gun Wagon II," the front of which was changed in the course of development, weighed about 11 tons in fighting trim, and carried a five-man crew. This vehicle, also designated "Device 803," was officially "*leichte Feldhaubitze* 18/2 auf *Fahrgestell Panzerkampfwagen* (9 (Sf) (Sd.Kfz. 124)."

From 1942 on, 682 of these vehicles were issued to the light batteries of the armored artillery units (Sf). The original evocative name "Wespe" (Wasp) was dropped at Hitler's order of February 27, 1944. The FAMO firm in Warsaw delivered the chassis until 1944. FAMO built another 158 units of a "Munitions-*Selbstfahrlafette* auf *Fahrgestell Panzerkampfwagen* II." This vehicle had the same body as the "Wespe," but carried no gun. A driver and two cannoneers were its crew. Loaded with 90 rounds of 10.5 cm ammunition, its total weight was about 11 tons. After rebuilding, the vehicles could also be used as gun carriers.

The "*Geschützwagen* II," which had taken over the unchanged "Panzer II" chassis in the beginning, was lengthened somewhat to the rear during its development by moving the leading wheels backward. The last pair of road wheels was also supported by truncated-cone springs. In February 1943 Hitler praised the light self-propelled mount on Panzer II as a very good solution. In order to meet the extraordinary need for these vehicles and, above all, to simplify types quickly, Speer's suggestion to switch the entire

Panzer II production, which was to be increased for a short time to 100 units per month, to light self-propelled gun mounts was accepted. Under these conditions it was also possible to send a suitable number of ammunition carriers to the troops. On March 6, 1943, Hitler again canceled Speer's suggestion that motors, gearboxes, and other components only be taken for self-propelled mounts when there were no more tank hulls at hand to use them. Hitler also required the investigation of the possibility of using the Panzer II production capacity that was being used for self-propelled mounts of light assault guns.

In April 1943, Hitler agreed to use the capacity for Panzer II building at FAMO-Ursus to build a tank destroyer exclusively for the armored troops. The 7.5 cm Pak L/48 was planned for use as its gun.

Production of the light howitzer mounts, as already noted, was being done mainly by FAMO in Warsaw, but was not switched to assault guns during the course of the war.

"Panzer II" chassis of Types A, B, and C had already been used in small numbers since 1941 as self-propelled mounts for the Heavy Infantry Gun 33 after removing the upper box body. Since a similar use of the "Panzer I" chassis was unsatisfactory, a body with a comparatively low height was now used.

With a five-man crew and a total weight of about 12 tons, these "*Geschützwagen* II *für* 15 cm sIG 33" were used mainly by Armored Grenadier units.

The "LaS 100" chassis was used to make the "*Selbstfahrlafette* II *für* 7.5 cm Pak 40/2" (Sd.Kfz. 131).

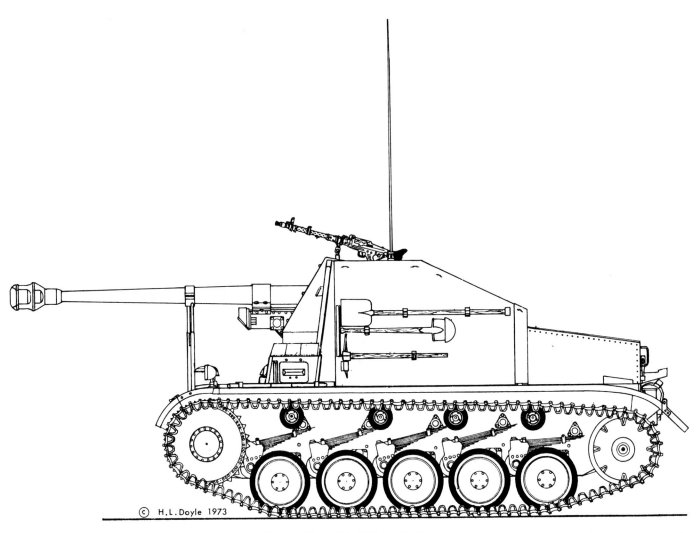

Sfl II for 7.5 cm Pak 40/2 (Sd.Kfz. 131)

Above: More than 1200 of these vehicles were issued to antitank units. Below: This picture shows the installation of the 7.5 cm Pak 40/2 in the "Marder II." Note the doubled armor of the gun shield.

Since the acceptance of these guns strongly limited the use of normal "Panzer II" chassis, the chassis was made wider experimentally, and lengthened with six-wheel running gear. Here, too, the insufficient engine power made the introduction of these vehicles impossible. Only prototypes were built.

Armored command cars on the "Panzer II" chassis differed from the tanks only by having dummy armament and more radio equipment. The armored artillery also received some fire-control tanks on "Panzer II" chassis. These vehicles were recognizable by a frame antenna over the engine compartment.

The Wegmann firm in Kassel had received a contract on October 18, 1939, to use the "LaS 100" chassis to build a minesweeping vehicle. This so-called "*Hammerschlag* Device," which was built onto the front of the "Panzer II," weighed some 2.5 tons, and was supposed to dispose of tellermines. Three test models were actually built.

In an official order of 1940, the existence of a "*Panzerkampfwagen* II" (Bridgelayer) was first mentioned. The Magirus Works in Ulm actually equipped a "b" chassis with a folding, extendable bridge that was suitable for crossing small

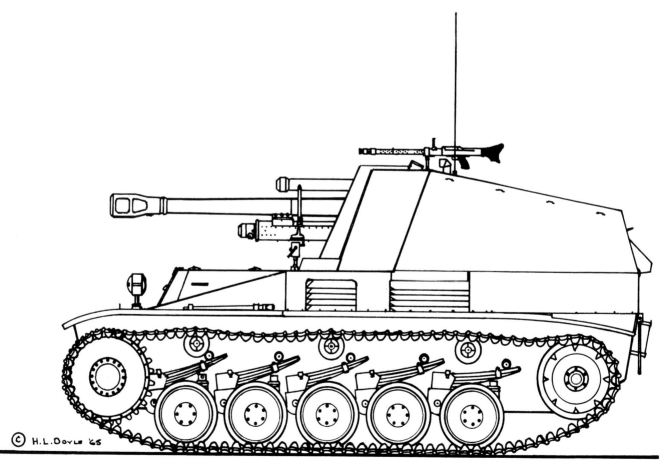

Leichte Panzerhaubitze 18/2 auf Fgst. Pz Kgfwg II (Sf) (Sd.Kfz. 124).

rivers. In troop use, though, this version had not proven itself. The HM 1941 No. 117 of January 23, 1941, states officially that the bridgelaying platoon formerly intended for armored engineer companies, according to former plans, was to be dropped, since appropriate vehicles were not available, and could not be built in a short time. A new establishment of the bridgelaying platoon as the third platoon of Bridge Column K was supposed to be ordered separately at the given time.

A number of "Panzer II" vehicles—without turrets—were assigned to the armored engineer companies from 1942 on. They were fitted with wooden load-carrying bodies that were covered by tent canvas. These vehicles were designated "*Pionier-Kampfwagen* II."

One of the "VK 1601" vehicles had been rebuilt as a recovery tank, with only a crane added in place of the turret. The vehicle was captured by the Allies in Italy.

At the beginning of assault-gun development, some "Panzer II" chassis were used to train tank drivers, as well as for use as makeshift mounts for antitank guns.

In all, it can be said that the "Panzer I and II" created to help in training, allowed the building up of the German armored weapon, and essentially supported the success gained at the beginning of the war.

The "Light Field Howitzer 18/2 on *Panzerkampfwagen* II chassis (Sf)" (Sd.Kfz. 124). These "Gun Wagon II" vehicles, also known by the name of "Wespe" (Wasp), were useful support weapons for the armored troops.

Front and rear views of the "Wespe" vehicle.

This top view shows the open-top fighting compartment of these vehicles. In the gun car, the motor was moved forward. The driver's front area had been changed several times in the course of this vehicle's development.

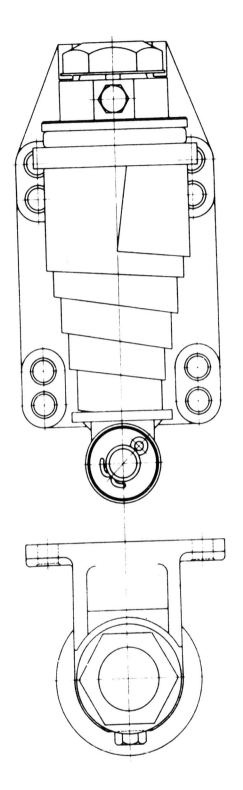
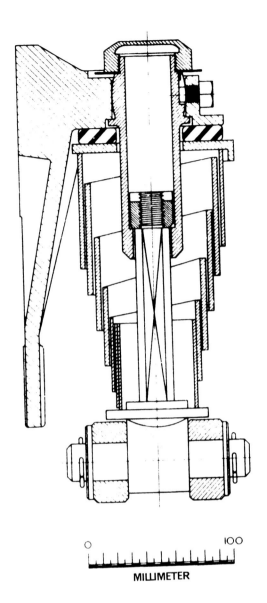

To strengthen the chassis, especially when firing, additional truncated-cone springs were installed.

158 of this "self-propelled ammunition carrier on Panzer II chassis" were built by FAMO. They were used to supply ammunition to "Wespe" units.

As a support weapon for armored grenadier units, these 15 cm Heavy Infantry Gun 33 self-propelled mounts were built on "Panzer II" chassis.

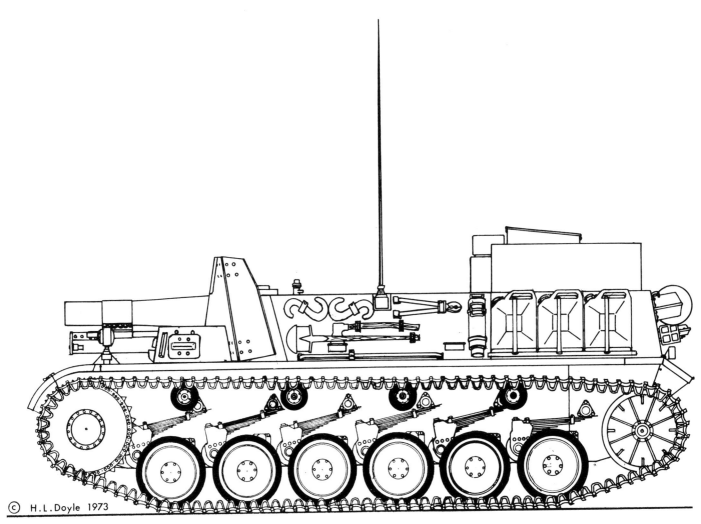

15 cm sIG (Sf) auf Fgst. Pz. II.

Above: A widened and lengthened "Panzer II" chassis was used experimentally to carry the 15 cm infantry gun. The engine power proved to be insufficient.
Below: A few "Panzer II" with added radio equipment appeared as observation-fire control vehicles of the armored artillery.

Armored engineer units received rebuilt "Panzer II" chassis as "*Pionier-Panzerwagen.*"

This recovery tank was a lone example of a rebuilding on the "VK 1601" chassis. The vehicle was captured by the Allies in Italy.

The Magirus firm mounted a fast bridge to cross small terrain obstacles on a "b" chassis. These bridgelaying tanks were not introduced; only this prototype was built.

Technical Data

Grosstraktor I

Type: "G.T.I"
Made in Germany
Made by: Daimler-Benz AG, Stuttgart-Untertürkheim (assembled at Unterlüss)

Years built: 1927-1929
Information source: Daimler-Benz archives
Notes: Contract given for two vehicles on March 26, 1927. Two vehicles built.

Motor	Mercedes-Daimler "DIVb"*
Cylinders	6, in-line
Bore	182 mm
Stroke	200 mm
Displacement	31,200 cc
Compression ratio	4.5:1
Engine speed	1450 rpm
Engine power	260/330 HP
Power to weight	20 HP/ton
Valves	drop
Crankshaft bearings	
Carburetors	2 Mercedes-Pallas
Firing order	1-5-3-6-2-4
Starter	DKW 2-stroke 2-cylinder, 10 HP at 3000 rpm
Generator	Bosch
Batteries	
Fuel supply	Pump
Cooling	Water
Clutch	Hydraulic outer band
Gearbox	Hydraulic planetary
Gears	3 forward, 1 reverse, + auxiliary
Drive wheels	rear
Axle ratio	
Top speed	40 km/h, 4 km/h in water**
Range	
Steering	hydraulic
Turning circle	
Suspension	leaf, longitudinal for trucks
Lubrication	central
Brakes made by	A. Teves, Daimler-Benz
Brake effect	hydraulic
Brake type	outer band
Footbrake works on	gearbox
Handbrake works on	left steering brake
Wheels	road wheels (300) 90-203, jack rollers
Vehicle track	2220 mm
Track on ground	6000 mm from leading to drive wheel
Track width	380 mm, chain type MK 6/380/160
Ground clearance	
Overall length	
Overall width	
Overall height	
Ground pressure	
Chassis weight	4193 kg, self-bearing box
Gross weight	15,000 kg
Load limit	
Seats	6
Fuel consumption	
Oil consumption	Depends on action
Fuel capacity	
Armor: turret	soft steel 14.5 mm
	hull soft steel 14,5 mm
Upgrade	30 degrees
Armament	1 7.5 Cannon L/20, 2-3 MG
Uses	Test model for tanks

* World War I aircraft motor, Type "F 182 206," weight with starter motor ca. 600 kg, motor length 1801 mm
** Two-propeller drive in water.

Panzerkampfwagen I (MG) (Sd.Kfz. 101) Type A

Type: "I A LaS Krupp"
Made in Germany
Made by Friedrich Krupp AG, Essen etc.

Years built: 1933-34
Information source: D 650/1 of September 20, 1938
Notes: Chassis numbers 10 001 – 10 477

Motor	Krupp "M 305"
Cylinders	4 opposed
Bore	92 mm
Stroke	130 mm
Displacement	3460 cc
Compression ratio	5.2:1
Engine speed	2500 rpm
Engine power	57 HP
Power to weight	11.1 HP/ton
Valves	drop
Crankshaft bearings	2 journal
Carburetors	2 Solex 40 JPF
Firing order	1-3-4-2
Starter	SSW LAK 300 generator
Generator	see starter
Batteries	1 6-volt 105 Ah
Fuel supply	pump
Cooling	air, blower
Clutch	two-plate dry
Gearbox	ZF Aphon FG 35
Gears	5 forward, 1 reverse
Drive wheels	front
Axle ratio	1:13.1 (intermediate)
Top speed	37 km/h
Range	road 145, off 100 km
Steering	Krupp clutch
Turning circle	2.1 meters
Suspension	coil and 1/4
Lubrication	high-pressure
Brakes made by	Krupp
Brake effect	mechanical
Brake type	band brakes
Footbrake works on	steering brake
Handbrake works on	none
Wheels	road wheels 530 x 80, jack rollers 700 (190) 85-72 (39)
Vehicle track	1676 mm
Track on ground	2470 mm
Track width	280 mm
Ground clearance	295 mm
Length overall	4020 mm
Width overall	2060 mm
Height overall	1720 mm
Firing height	1500 mm
Ground pressure	0.39 kg/cm^2
Chassis weight	
Total weight	5400 kg
Load limit	340 kg
Seats	2
Fuel consumption	100 liters/100 km
Oil consumption	depends on use
Fuel capacity	144 liters, 2 tanks
Armor, hull & turret	13 mm all around
Performance:	
Upgrade	30 degrees
Climbing	370 mm
Wading	600 mm
Spanning	1400 mm
Armament	2 MG 13 (Dreyse) (1525)*
Uses	light training tank

* number in () after armament = rounds of ammunition carried

Panzerkampfwagen I (MG) (Sd.Kfz. 101) Type B

Type: "I B LaS May"
Made in Germany
Made by Henschel & Sohn GmbH, Kassell etc.

Years built: 1934-41
Information source: D 650/4 of February 23, 1938

Motor	Maybach "NL 38 TR"
Cylinders	6 in-line
Bore	90 mm
Stroke	100 mm
Displacement	3791 cc
Compression ratio	6.7:1
Engine speed	3000 rpm
Engine power	100 HP
Power to weight	17.1 HP/ton
Valves	drop
Crankshaft bearings	8 journal
Carburetors	1 Solex 40 JFF II
Firing order	1-5-3-6-2-4
Starter	Bosch BJH 1.4/12 ARS 113
Generator	Bosch GTL 600/12-1200 RS 39
Batteries	1 12-volt, 105 Ah
Fuel supply	Pallas pump Type C, 949
Cooling	water
Clutch	Two-plate dry
Gearbox	ZF Aphon FG 31
Gears	5 forward, 1 reverse
Drive wheels	front
Axle ratio	
Top speed	40 km/h
Range	road 140, off 115 km
Steering	Krupp clutch
Turning circle	2.1 meters
Suspension	coil and 1/4
Lubrication	high-pressure
Brakes made by	Friedrich Krupp
Brake effect	mechanical
Brake type	band
Footbrake works on	steering brake
Handbrake works on	none
Wheels	road wheels 530 x 80, jack rollers 700 (190) 85-72
Vehicle track	1670 mm
Track on ground	2470 mm
Track width	280 mm
Ground clearance	295 mm
Overall length	4420 mm
Overall width	2060 mm
Overall height	1720 mm
Firing height	1500 mm
Ground pressure	0.43 kg/cm^2
Chassis weight	5660 kg of its own
Total weight	6000 kg
Load limit	340 kg
Seats	2
Fuel consumption	road 100, off 125 liters/100 km
Oil consumption	depends on action
Fuel capacity	146 liters
Armor hood & turret	13 mm all around
Upgrade	30 degrees
Climbing	370 mm
Wading	600 mm
Spanning	1400 mm
Armament	2 MG 13 (Dreyse) 2250 rounds
Uses	light training tank

Panzerkampfwagen I (Type C) (VK 601)

Type: "VK 601"
Made in Germany
Made by Krauss-Maffei AG, Munich-Allach

Years built: 1939-1941
Information source: D 650/22 of 10/1/1942 & KM archives
Notes: Order for 40 units on 9/15/1939, 40 built

Motor	Maybach "HL 45 p"
Cylinders	6 in-line
Bore	95 mm
Stroke	110 mm
Displacement	4678 cc
Compression ratio	6.6:1
Egine speed	3800 rpm
Engine power	150 HP
Power to weight	18.2 HP/ton
Valves	drop
Crankshaft bearings	8 journal
Carburetors	2 Solex 40 JFF II
Firing order	1-5-3-6-2-4
Starter	Bosch BNG
Generator	Bosch RKCN 300/12-1300
Batteries	2 12-volt, 105 Ah
Fuel supply	pump
Cooling	water
Clutch	multiplate in oil
Gearbox	Maybach "VG 15319"
Gears	8 forward, 2 reverse
Drive wheels	front
Axle ratio	1:5.5, intermediate gear
Top speed	79 km/h
Range	300 km
Steering	"KM LG 45 R" 3 radial link
Turning circle	on the spot
Suspension	torsion bars, transverse
Lubrication	high-pressure
Brakes made by	Krauss-Maffei
Brake effect	mechanical
Prake type	outside shoe
Footbrake works on	drive, intermediate gear
Handbrake works on	drive
Wheels	steel discs & spokes
Vehicle track	1630 mm
Track on ground	2200, 89 links
Track width	390 mm
Ground clearance	290 mm
Overall length	4195 mm
Overall width	1920 mm
Overall height	2010 mm
Firing height	1724 mm
Ground pressure	0.84 kg/cm^2
Chassis weight	
Total weight	8000 kg
Load limit	1500 kg
Seats	2
Fuel consumption	
Oil consumption	depends on action
Armor: hull	10-30 mm
Upgrade	
Climbing	300 mm
Wading	785 mm
Spanning	1200 mm
Armament	1 EW 141 + 1 MG 34
Uses	Prototype for fast reconnaissance car and for airborne troop use

Panzerkampfwagen II (Type F) (VK 1801)

Type: "VK 1801"
Made in Germany
Made by Krauss-Maffei AG, Munich-Allach

Years built: 1939-40
Information source: D 650/33 of July 1942 & KM archives
Notes: Contract of 12/22/1939, 30 units built

Motor	Maybach "HL 45 p"
Cylinders	6 in-line
Bore	95 mm
Stroke	110 mm
Displacement	4678 cc
Compression ratio	6.6:1
Engine speed	3800 rpm
Enguine power	150 HP
Power to weight	7.1 ps/ton
Valves	drop
Crankshaft bearings	8 journal
Carburetors	2 Solex 40 JFF II
Firing order	1-5-3-6-2-4
Starter	Bosch BNG 2.5/12
Generator	Bosch RKCN 300/12-1300
Batteries	2 12-volt, 105 Ah
Fuel supply	pump
Cooling	water
Clutch	multidisc dry
Gearbox	ZF "SSG 47"
Gears	4 forward, 1 reverse
Drive wheels	front
Axle ratio	
Top speed	25 km/h
Range	150 km
Steering	clutch
Turning circle	2.10 meters
Suspension	torsion bars, transverse
Lubrication	high-pressure
Brakes made by	Krauss-Maffei
Brake effect	mechanical
Brake type	outside shoes
Footbrake works on	drive, intermediate gear
Handbrake works on	drive
Wheels	steel plate discs & spokes
Vehicle track	2100 mm
Track on ground	200 mm, 52 links per track
Track width	540 mm
Ground clearance	350 mm
Ground clearance	350 mm
Overall length	4375 mm
Overall width	2640 mm
Overall height	2050 mm
Firing height	1750 mm
Ground pressure	0.46 kg/cm^2
Chassis weight	
Total weight	21,000 kg
Load limit	1500 kg
Seats	2
Fuel consumption	
Oil consumption	depends on action
Armor: hull & turret: 80 mm	
Upgrade	59 %
Climbing	330 mm
Wading	570 mm
Spanning	1500 mm
Armament	2 MG 34
Uses	heavily armored infantry support

Panzerkampfwagen I (A) Ammunition Tractor (Sd.Kfz. 111)

Type: "I A LaS Krupp"
Made in Germany
Made by Daimler-Benz AG, Berlin-Marienfelde

Years built: 1934-35
Information source: D 650/9b of 8/15/1942
Notes: 51 units rebuilt

Motor	Krupp "M 305"
Cylinders	4 opposed
Bore	92 mm
Stroke	139 mm
Displacement	3460 cc
Compression ratio	5.2:1
Engine speed	2500 rpm
Engine power	57 HP
Power to weight	12 HP/ton
Valves	drop
Crankshaft bearings	2 journal
Carburetors	2 Solex 40 JFP
Firing order	1-3-4-2
Starter	SSW generator LAK 300
Generator	see starter
Batteries	1 6-volt, 105 Ah
Fuel supply	pump
Cooling	air, blower
Clutch	2-plate dry
Gearbox	ZF Aphon FG 35
Gears	5 forward, 1 reverse
Drive wheels	front
Axle ratio	1:1.31 (intermediate gear)
Top speed	37 km/h
Range	road 140, off 95 km
Steering	Krupp clutch
Turning circle	2.1 meters
Suspension	screw and 1/4
Lubrication	high-pressure
Brakes made by	Krupp
Brake effect	mechanical
Brake type	band
Footbrake works on	steering brake
Handbrake works on	none
Wheels	road wheels 530 x 80 + jack rollers 700 (190) 85-72 (39)
Vehicle track	1676 mm
Track on ground	2470 mm
Track width	280 mm
Ground clearance	295 mm
Overall length	4020 mm
Overall width	2060 mm
Overall height	1400 mm
Ground pressure	0.35 kg/cm^2
Chassis weight	
Total weight	5000 kg
Load limit	500 kg
Seats	2
Fuel consumption	100 liters/100 km
Oil consumption	depends on action
Fuel capacity	144 liters, 2 tanks
Armor: hull front	15 mm
hull sides	13 mm
Upgrade	30 degrees
Climbing	370 mm
Wading	600 mm
Spanning	1400 mm
Uses	armored ammunition carrier

4.7-cm-Pak (t) auf *Panzerkampfwagen* I (Sd.Kfz. 101) without turret

Type: "I B LaS May"
Made in Germany
Made by: Altmärkische Kettenfabrik GmbH, Berlin-Spandau (rebuilding)

Years built: 1939-40
Information source: D 650/17 of May 15, 1940
Notes: 132 units rebuilt

Motor	Maybach/Nordbau "NL 38 TR"
Cylinders	6 in-line
Bore	90 mm
Stroke	100 mm
Displacement	3791 cc
Compression ratio	6.7:1
Engine speed	3000 rpm
Engine power	100 HP
Power to weight	15.6 HP/ton
Valves	drop
Crankshaft bearings	8 journal
Carburetors	1 Solex 40 JFF II
Firing order	1-5-3-6-2-4
Starter	Bosch BJH 1,4/12
Generator	Bosch GTL 600/12-1200
Batteries	2 12-volt 105 Ah
Fuel supply	Pallas pump C.949
Cooling	water
Clutch	2-plate dry
Gearbox	ZF Aphon FG 31
Gears	5 forward, 1 reverse
Drive wheels	front
Axle ratio	
Top speed	42 km/h
Range	road 140, off 95 km
Steering	Krupp clutch
Turning circle	2.1 meters
Suspension	coil and 1/4
Lubrication	high-pressure
Brakes made by	Friedrich Krupp
Brake effect	mechanical
Footbrake works on	steering brake
Handbrake works on	none
Wheels	road wheels 530 x 80, jack rollers 700 (190) 85-72
Vehicle track	1670 mm
Track on ground	2470 mm
Track width	280 mm
Ground clearance	295 mm
Overall length	4420 mm
Overall width	1850 mm
Overall height	2250 mm
Firing height	1720 mm
Ground pressure	0.45 kg/cm^2
Chassis weight	
Total weight	6400 kg
Load limit	300 kg
Seats	3
Fuel consumption	road 100, off 125 liters/100 km
Oil consumption	depends on action
Fuel capacity	146 litters, 2 tanks
Armor: hull:	13 mm all around
body front	14.5 mm
body sides	14.5 mm
Upgrade	30 degrees
Climbing	370 mm
Wading	600 mm
Spanning	1400 mm
Armament	1 4.7 cm Pak (t) (86 rounds) & 1 MP (192
Uses	light makeshift tank destroyer

Geschützwagen I für 15-cm-sIG 33

Type: "I B LaS May"
Made in Germany
Made by Altmärkische Kettenfabrik GmbH, Berlin-Spandau (rebuilding)

Years built: 1939-40
Information source: D 650/4 of February 23, 1938
Notes: 38 vehicles created through rebuilding

Motor:	Maybach "NL 38 TR"
Cylinders	6 in-line
Bore	90 mm
Stroke	100 mm
Displacement	3791 cc
Compression ratio	6.7:1
Engine speed	3000 rpm
Engine power	100 hp
Power to weight	11.9 HP/ton
Valves	drop
Crankshaft bearings	8 journal
Carburetors	1 Solex 40 JFF II
Firing order	1-5-3-6-2-4
Starter	Bosch BJH 1.4/12
Generator	Bosch GTL 600/12-1200
Batteries	2 12-volt, 105 Ah
Fuel supply	Pallas pump C.949
Cooling	water
Clutch	two-plate dry
Gearbox	ZF "FG 31" Aphon
Gears	5 forward, 1 reverse
Drive wheels	front
Axle ratio	
Top speed	35 km/h
Range	100 km
Steering	Krupp clutch
Turning circle	2.1 meters
Suspension	coil and 1/4
Lubrication	high-pressure
Brakes made by	Friedrich Krupp AG
Brake effect	mechanical
Brake type	band
Footbrake works on	steering brake
Handbrake works on	none
Wheels	road wheels 530 x 80, jack rollers 700 (190) 85-72
Vehicle track	1670 mm
Track on ground	2470 mm
Track width	280 mm
Ground clearance	295 mm
Overall length	4420 mm
Overall width	2680 mm
Overall height	3350 mm
Ground pressure	0.6 kg/cm^2
Chassis weight	
Total weight	8500 kg
Load limit	2000 kg
Seats	4
Fuel consumption	125 liters/100 km
Oil consumption	depends on action
Fuel capacity	146 liters, 2 tanks
Armor: hull	13 mm all around
body	10 mm all around
Upgrade	20 degrees
Climbing	370 mm
Wading	600 mm
Spanning	1400 mm
Armament	1 15 cm sIG 33 L/11
Uses	makeshift infantry support gun

Heavy Armored Command Car (Sd.Kfz.265)

Type: "kl. B"
Manufacturer: Daimler-Benz AG, Berlin-Marienfelde Works

Years built: 1936-38
Sources of information: D 650/10 of October 25, 1938
Notes: circa 200 built
Chassis numbers: 15 001-15 200

Motor	Maybach "NL 38 TR"
Cylinders	6 in-line
Bore	90 mm
Stroke	100 mm
Displacement	3791 cc
Compression ratio	6.7:1
Engine speed	3000 rpm
Engine power	100 HP
Power to weight	17 HP/ton
Valves	drop
Crankshaft bearings	8 journal
Carburetors	1 Solex 40 JFF 2 F
Firing order	1-5-3-6-2-4
Starter	Bosch BJH 1,4/12 ARS 113
Generator	Bosch GTL 600/12-1200
Batteries	2 12-volt, 105 Ah
Fuel supply	Pallas pump 75032 Type C.949
Cooling	water
Clutch	two-plate dry
Gearbox	ZF Aphon FG 31
Gears	5 forward, 1 reverse
Drive wheels	front
Axle ratio	
Top speed	40 km/h
Range	road 170, off 115 km
Steering	Krupp clutch
Turning circle	2.1 meters
Suspension	high-pressure
Brakes made by	Krupp
Brake effect	mechanical
Brake type	band
Footbrake works on	steering brake
Handbrake works on	none
Wheels	road wheels 530 x 80, jack rollers 700 (190) 85-72 (39)
Vehicle track	1670 mm
Track on ground	2400 mm
Track width	280 mm
Ground clearance	295 mm
Overall length	4420 mm
Overall width	2060 mm
Overall height	1990 mm
Firing height	1480 mm
Ground pressure	0.43 kg/cm^2
Chassis weight	
Total weight	5880 kg
Load limit	450 kg
Seats	3
Fuel consumption	110 liters/100 km
Oil consumption	depends on action
Fuel capacity	146 liters
Armor: hull	14,5 mm all around
Upgrade	30 degrees
Climbing	370 mm
Wading	600 mm
Spanning	1400 mm
Armament	1 MG 34 (900 rounds)
Uses	light command car for armored units

Panzerkampfwagen II (2 cm) 9Sd.Kfz. 121) Ausf. A1, a2 und a3

Type: "1/LaS 100"
Made in Germany
Made by Maschinenfabrik Augsburg-Nürnberg AG. Nürnberg works

Years built: 1934-36
Information source: D 651/1 of March 31, 1938
Notes: chassis 20 001-20 010, 20 011-20 025, 20 026-20 050, 20 051-21 000

Motor	Maybach "HL 57 TR"	Brakes made by	MAN
Cylinders	6 in-line	Brake effect	mechanical
Bore	100 mm	Brake type	inner/outer shoes
Stroke	120 mm	Footbrake works on	steering brake
Displacement	5698 cc	Handbrake works on	steering brake
Compression ratio	6.3:1	Wheels	road wheels & jack rollers
Engine speed	2600 rpm	Vehicle track	1780 mm
Engine power	130 HP	Track on ground	2426 mm, 108 links per track
Power to weight	17 HP/ton	Track width	300 mm
Valves	drop	Ground clearance	300 mm
Crankshaft bearings	8 journal	Overall length	4380 mm
Carburetors	1 Solex 40 JFF II	Overall width	2140 mm
Firing order	1-5-3-6-2-4	Overall height	1945 mm
Starter	Bosch BNF 2.5/12 BRS 112+AL/ZMA/R 8	Ground pressure	0.5 kg/cm^2
Generator	Bosch RKC 130/12-825 LS 44	Chassis weight	5200 kg
Batteries	1 or 2 12-volt, 105 or 60 Ah	Total weight	7600 kg
Fuel supply	pump	Load limit	
Cooling	water	Seats	3
Clutch	Two-plate dry	Fuel consumption	150 liters/100 km
Gearbox	ZF SSG 45 (2nd-6th synchromesh)	Oil consumption	depends on action
Gears	6 forward, 1 reverse	Fuel capacity	102 + 68 = 170 liters
Drive wheels	front	Armor: hull	14.5 mm all around
Axle ratio	1.9:1 (intermediate gear)	Upgrade	30 degrees
Top speed	40 km/h	Climbing	420 mm
Range	road 210, off 160 km	Wading	920 mm
Steering	MAN planetary	Spanning	1800 mm
Turning circle	4.8 meters	Armament	1 2 cm KwK 30 (180 rounds) + 1 MG 34 (1425 rounds)
Suspension	leaf springs, longitudinal, paired road wheels		
Lubrication	high-pressure	Uses	light tank

Panzerkampfwagen II (2 cm) (Sd.Kfz. 121) Ausf. B

Made in Germany
Made by Maschinenfabrik Augsburg-Nürnberg AG, Nürnberg works, etc.
Type: "2/LaS 100"

Years built: 1936-37
Information source: D 651/1 of 3/31/1938

Notes: chassis no. 21 001-21 100

Motor	Maybach "HL 62 TR"
Cylinders	6 in-line
Bore	105 mm
Stroke	120 mm
Displacement	6191 cc
Compression ratio	6.5:1
Engine speed	2600 rpm
Engine power	140 HP
Power to weight	17.5 HP/ton
Valves	drop
Crankshaft bearings	8 journal
Carburetors	1 Solex 40 JFF II
Firing order	1-5-3-6-2-4
Starter	Bosch BNG 2.5/12 BR 183+AL/ZMA/R 3
Generator	Bosch RJJK 130/12-1500
Batteries	1 12-volt 105 Ah
Fuel supply	pump
Cooling	water
Clutch	2-plate dry F&S K230K
Gearbox	ZF SSG 45 (2nd-6th synchromesh)
Gears	6 forward, 1 reverse
Drive wheels	front
Axle ratio	1:9.1 (intermediate gear)
Top speed	40 km/h
Range	road 190, off 125 km
Steering	MAN planetary with intermediate gear
Turning circle	4.8 meters
Suspension	leaf springs, longitudinal, road wheels paired
Lubrication	high-pressure
Brakes made by	MAN
Brake effect	mechanical
Brake type	shoe
Footbrake works on	steering brakes
Handbrake works on	Steering brakes
Wheels	Road wheels & jack rollers
Vehicle track	1780 mm
Track on ground	2418 mm
Track width	300 mm
Ground clearance	312 mm
Overall length	4755 mm
Overall width	2140 mm
Overall height	1955 mm
Ground pressure	0.56 kg/cm^2
Chassis weight	5500 kg
Total weight	7900 kg
Load limit	
Seats	3
Fuel consumption	150 liters/100 km
Oil consumption	depends on action
Fuel capacity	102 + 68 = 170 liters
Armor: hull	14.5 mm all around
Turret	14.5 mm all around
Upgrade	30 degrees
Climbing	630 mm
Wading	892 mm
Spanning	1800 mm
Armament	1 2 cm KwK 30 (180 rounds) + 1 MG 34 (1425 rounds)
Uses	light tank

Panzerkampfwagen II (2 cm) (Sd.Kfz. 121) Ausf. C

Made in Germany
Made by Maschinenfabrik Augsburg-Nürnberg AG, Nürnberg works etc.
Type: "3/LaS 100"

Year built: 1937
Information source: D 651/1 of 3/31/1937

Notes: Chassis no. 21 101-22 000, 22 001-23 000

Motor	Maybach "HL 62 TR"
Cylinders	6 in-line
Bore	105 mm
Stroke	120 mm
Displacement	6191 cc
Compression ratio	6.5:1
Engine speed	2600 rpm
Engine power	140 HP
Power to weight	16 HP/ton
Valves	drop
Crankshaft bearings	8 journal
Carburetors	1 Solex 40 JFF II
Firing order	1-5-3-6-2-4
Starter	Bosch BNG 2.5/12 BR 183+AL/ZMA/R 3
Generator	Bosch GTLN 600/12-1500
Batteries	1 12-volt, 105 Ah
Fuel supply	Pallas pump
Cooling	water
Clutch	two-plate dry F&S K230K
Gearbox	ZF SSG 45 (2nd-6th synchromesh)
Gears	6 forward, 1 reverse
Drive wheels	front
Axle ratio	1:9/1 (intermediate gear)
Top speed	40 km/h
Range	road 190, off 125 km
Steering	MAN planetary with intermediate gear
Turning circle	4.8 meters
Suspension	leaf springs, longitudinal, wheels single
Lubrication	high-pressure
Brakes made by	MAN
Brake effect	mechanical
Brake type	outer band
Footbrake works on	steering brakes
Handbrake works on	steering brakes
Wheels	road wheels 550 x 100-55, jack rollers 220 x 105
Vehicle track	1880 mm
Track on ground	2400 meters, 108 links per track
Track width	300 mm
Ground clearance	345 mm
Overall length	4810 mm
Overall width	2223 mm
Overall height	1990 mm
Ground pressure	0.56 kg/cm^2
Chassis weight	6500 kg
Total weight	8900 kg
Load limit	
Seats	3
Fuel consumption	150 liters/100 km
Oil consumption	depends on action
Fuel capacity	102 + 68 = 170 liters
Armor: hull	14.5 mm all around
turret	14.5 mm all around
Upgrade	30 degrees
Climbing	630 mm
Wading	925 mm
Spanning	1800 mm
Armament	1 2 cm KwK 30 (180 rounds) + 1 MG 34 (1425 rounds)
Uses	light tank

Panzerkampfwagen II (2 cm) (Sd.Kfz. 121) Ausf. A, B, C und F

Made in Germany
Made by Maschinenfabrik Augsburg-Nürnberg AG, Nürnberg works etc.
Type: "4 to 7/LaS 100"

Years made: 1937-1940
Price: RM 49,228 (without weapons)

Information source: D 651/1 of 3/31/1938
Notes: chassis 23 001–24 000 = A, 24 001-26 000 = B. 26 001-27 000 = C, 28 001-29 400 = F

Motor	Maybach/Nordbau "HL 62 TRM"
Cylinders	6 in-line
Bore	105 mm
Stroke	120 mm
Displacement	6191 cc
Compression ratio	6.5:1
Engine speed	2600 rpm
Engine Power	140 HP
Power to weight	14.75 HP/ton
Valves	drop
Crankshaft bearings	8 journal
Carburetors	1 Solex 40 JFF II
Firing order	1-5-3-6-2-4
Starter	Bosch BNG 2.5/12+AL/ZMA
Generator	Bosch GTLN 600/12-1500
Batteries	1 12-volt, 120 Ah
Fuel supply	Pallas pump no. 62601
Cooling	water
Clutch	two-plate dry F & S K 230 K
Gearbox	ZF SSG 46 Aphon
Gears	6 forward, 1 reverse
Drive wheels	front
Axle ratio	
Top speed	40 km/h
Range	road 190, off 125 km
Steering	MAN-Wilson clutch
Turning circle	4.8 meters
Suspension	leaf springs, longitudinal, single-wheel springs
Lubrication	high-pressure
Brakes made by	MAN
Brake effect	mechanical
Brake type	outer band, self-adjusting
Footbrake works on	Steering brakes
Handbrake works on	Steering brakes
Wheels	Road wheels 550 x 100-55, jack rollers 220 x 105
Vehicle track	1880 mm
Track on ground	2400 mm
Track width	300 mm
Ground clearance	345 mm
Overall length	4810 mm
Overall width	2280 mm
Overall height	2020 mm, Type F 2150 mm
Firing height	1595 mm
Ground pressure	0.76 kg/cm^2
Chassis weight	6800 kg
Total weight	9500 kg
Load limit	
Seats	3
Fuel consumption	150 liters/100 km
Oil consumption	depends on action
Fuel capacity	170 liters
Armor: hull front	14.5 + 20 mm
hull other	14.5 mm
turret front	14.5 + 14.5 + 20 mm
turret other	14.5 mm
Upgrade	30 degrees
Climbing	420 mm
Wading	925 mm
Spanning	1800 mm
Armament	1 2 cm KwK 30 (180 rounds) + 1 MG 34 (1425, belt feed 2100)
Uses	light tank

Panzerkampfwagen II (2 cm) (Sd.Kfz.121) Types D & E

Made in Germany
Made by Maschinenfabrik Augsburg-Nürnberg, Nürnberg works, etc.
Type: "8/LaS 138"

Years built: 1938-1939
Information source: D 651/11, 5/2/1939

Notes: chassis 27 001-27 800 = D, 27 801-28 000 = E

Motor	Maybach "HL 62 TRM"
Cylinders	6 in-line
Bore	105 mm
Stroke	120 mm
Displacement	6191 cc
Compression ratio	6.5:1
Engine speed	2600 rpm
Engine power	140 HP
Power to weight	12.8 HP/ton
Valves	drop
Crankshaft bearings	7 + 1 journal
Carburetors	1 Solex 40 JFF II
Firing order	1-5-3-6-2-4
Starter	Bosch BNG 2.5/12 + AL/ZMD/R 3
Generator	Bosch GTLN 600/12-1500
Batteries	1 12-volt, 120 Ah
Fuel supply	Pallas pump
Cooling	water
Clutch	two-plate dry F & S PF 20 K
Gearbox	Maybach Variorex VG 102128 H
Gears	7 forward, 1 reverse
Drive wheels	front
Axle ratio	1:5.9 (intermediate gear)
Top speed	55 km/h
Range	road 200, off 130 km
Steering	clutch, mechanical
Turning circle	
Suspension	torsion bars
Lubrication	high-pressure
Brakes made by	MAN
Brake effect	mechanical
Brake type	outer shoes
Footbrake works on	steering brake
Handbrake works on	drive
Wheels	pressed disc
Vehicle track	1920 mm
Track on ground	2200 mm, 96 links per track
Track width	300 mm
Ground clearance	290 mm
Overall length	4640 mm
Overall width	2300 mm
Overall height	2020 mm
Ground pressure	0.80 kg/cm^2
Chassis weight	
Total weight	10,000 kg
Load limit	1500 kg
Seats	3
Fuel consumption	road 100, off 150 liters/100 km
Oil consumption	0.3 liter/100 km
Fuel capacity	200 liters, 1 tank
Armor: hull front	30 mm
hull other	14.5 mm
angled	30 mm
sides + back	14.5 mm
Upgrade	24 degrees
Climbing	420 mm
Wading	850 mm
Spanning	1750 mm
Armament	1 2 cm KwK 30 (180 rounds) + 1 MG 34 (1425 rounds)
Uses	light tank, fast tank in "Light Divisions"

Panzerkampfwagen II (2 cm) (Sd.Kfz. 121) Types G1, G3, G4 and J

Made in Germany
Made by Maschinenfabrik Augsburg-Nürnberg AG, Nürnberg works, etc.
Type: "LaS 100"

Years built: 1940-42
Price: RM 49,229 (without weapons)

Information source: D 651/32 of 10/20/1942
Chassis 150 001-150 075 = G, 150 101-150 130 = J
Final versions of Panzer II

Motor	Maybach "HL 62 TRM"
Cylinders	6 in-line
Bore	105 mm
Stroke	120 mm
Displacement	6191 cc
Compression ratio	6.5:1
Engine speed	2600 rpm
Engine power	140 HP
Power to weight	14.75 HP/ton
Valves	drop
Crankshaft bearings	7 + 1 journal
Carburetor	1 Solex 40 JFF II
Firing order	1-5-3-6-2-4
Starter	Bosch BNG 2.5/12 + AL/ZMD/R 3
Generator	Bosch RJJK 130/12-1500
Batteries	1 12-volt, 120 Ah
Fuel supply	Pallas pump
Cooling	water
Clutch	two-plate dry F & S K 230 K
Gearbox	ZF SSG 46 Aphon
Gears	6 forward, 1 reverse
Drive wheels	front
Axle ratio	
Top speed	40 km/h
Range	road 190, off 125 km
Steering	MAN-Wilson clutch
Turning circle	4.8 meters
Suspension	leaf springs, longitudinal, single-wheel springs
Lubrication	high-pressure
Brakes made by	MAN
Brake effect	mechanical
Brake type	outer band, self-adjusting
Footbrake works on	steering brake
Handbrake works on	steering brake
Wheels	road wheels 550 x 98-455, jack rollers 20 x 105
Vehicle track	1880 mm
Track on ground	2400 mm, 108 links per track
Track width	300 mm
Overall length	4810 mm
Overall width	2280 mm
Overall height	2020 mm
Firing height	1595 mm
Ground pressure	0.76 kg/cm^2
Chassis weight	6800 kg
Total weight	9500 kg
Load limit	
Seats	3
Fuel consumption	road 90, off 135 liters/100 km
Oil consumption	0.3 liter/100 km
Fuel capacity	170 liter
Armor: hull front	35 mm
hull sides	20 mm
hull rear	14.5 mm
turret front	30 mm
turret other	14.5 mm
Upgrade	30 degrees
Climbing	420 mm
Wading	925 mm
Spanning	1700 mm
Armament	1 2 cm KwK 30 or 38 (180 rounds) + 1 MG 34 (2550 rounds)
Uses	light tank

Panzerkampfwagen II (2 cm KwK 28) (Sd.Kfz. 123) Type L (Luchs)

Made in Germany
Made by Maschinenfabrik Augsburg-Nürnberg AG, Nürnberg etc.
Type: "VK 1303"

Years built: 1942-43
Information source: Handbook WaA, Page K 45, etc.

Notes: chassis no. from 200 100, 131

Motor	Maybach "HL 66 P"
Cylinders	6 in-line
Bore	105 mm
Stroke	130 mm
Displacement	6754 cc
Compression ratio	6.5:1
Engine speed	2800/3200 max.
Engine power	180/200 HP
Power to weight	16.7 HP/ton
Valves	drop
Crankcase bearings	8 journal
Carburetor	2 Solex 40 JFF II
Firing order	1-5-3-6-2-4
Starter	Bosch BNG 2.5/12 BRS 161
Generator	Bosch GTN 600/12-1200 A 4
Batteries	1 12-volt, 120 Ah
Fuel supply	pump
Cooling	water
Clutch	two-plate dry F & S "Mecanü
Gearbox	ZF Aphon SSG 48
Gears	6 forward, 1 reverse
Drive wheels	front
Axle ratio	
Top speed	road 60, off 30 km/h
Range	road 290, off 175 km
Steering	MAN clutch
Turning circle	on the spot
Suspension	torsion bars, transverse
Lubrication	high-pressure
Brakes made by	MAN
Brake effect	mechanical
Brake type	outer shoes
Footbrake works on	drive wheels
Handbrake works on	drive wheels
Wheels	box layout
Vehicle track	2080 mm
Track on ground	2200 mm
Track width	360 mm
Ground clearance	400 mm
Overall length	4630 mm
Overall width	2480 mm
Overall height	2210 mm
Ground pressure	0.98 kg/cm^2
Chassis weight	
Total weight	11,800 kg
Load limit	1500 kg
Seats	4
Fuel consumption	road 90, off 150 liters
Oil consumption	depends on action
Fuel capacity	235 liters
Armor: hull front	30 mm
hull other	20 mm
turret front	30 mm
turret other	20 mm
Upgrade	30 degrees
Climbing	600 mm
Wading	1400 mm
Spanning	1600 mm
Armament	1 2 cm KwK 38 (320 rounds), as of no. 101 5 cm KwK, 39 L/60 + 1 MG 34 (2280 rounds)
Uses	armored scout car (full-track)

Panzerkampfwagen II (F) (Sd.Kfz. 122) Ausf. A und B

Made in Germany
Made by Wegmann & Co., Kassel (rebuilding)
Type: "8/LaS 138"

Year built: 1940
Information source: D 651/61 of 10/9/1941
Notes: Originally Panzer II Types D + E, 87 + 3 rebuilt

Motor	Maybach "HL 62 TRM"
Cylinders	6 in-line
Bore	105 mm
Stroke	120 mm
Displacement	6191 cc
Compression ratio	6.5:1
Engine speed	2600 rpm
Engine power	140 HP
Power to weight	12.7 HP/ton
Valves	drop
Crankshaft bearings	7 + 1 journal
Carburetor	1 Solex 40 JFF II
Firing order	1-5-3-6-2-4
Starter	Bosch BNG 2.5/12 + AL/ZMD/R 3
Generator	Bosch GTLN 600/12-1500
Batteries	1 12-volt, 120 Ah
Fuel supply	Pallas pump
Cooling	water
Clutch	two-plate dry F & S PF 20 K
Gearbox	Maybach Variorex VG 102128 H
Gears	7 forward, 3 reverse
Drive wheels	front
Axle ratio	1:5.9 (intermediate gear)
Top speed	40 km/h
Range	road 190, off 125 km
Steering	MAN clutch, mechanical
Turning circle	
Suspension	torsion bars
Lubrication	high-pressure
Brakes made by	MAN
Brake effect	mechanical
Brake type	outer shoes
Footbrake works on	steering brake
Handbrake works on	drive
Wheels	pressed disc wheels
Vehicle track	1920 mm
Track on ground	2200 mm, 96 links per track
Track width	300 mm
Ground clearance	290 mm
Overall length	4750 mm
Overall width	2140 mm
Overall height	1850 mm
Ground pressure	0.85 kg/cm^2
Chassis weight	
Total weight	11,000 kg
Load limit	2000 kg
Seats	2
Fuel consumption	road 100, off 150 liters/100 km
Oil consumption	0.3 liter/100 km
Fuel capacity	200 liters /100 km
Armor: hull front	30 mm
hull other	14,5 mm
Upgrade	30 degrees
Climbing	420 mm
Wading	800 mm
Spanning	1700 mm
Armament	2 flamethrowers (180 degrees) + 1 MG 34 (2000 rounds)
Uses	flamethrowing tank for special units

Pz.Sfl. 1 für 7.62-cm-Pak (Fahrgestell PzKgfwg. II Ausf. D2 (Sd.Kfz. 132)

Made in Germany
Made by Altmärkische Kettenfabrik GmbH, Berlin-Borsigwalde (rebuilding)
Type: "LaS 138"

Years built: 1942-44
Information source: D 651/13 of 4/27/1942

Notes: 210 units rebuilt

Motor	Maybach "HL 62 TRM"
Cylinders	6 in-line
Bore	105 mm
Stroke	120 mm
Displacement	6191 cc
Compression ratio	6.5:1
Engine speed	2600 rpm
Engine power	140 HP
Power to weight	12.2 HP/ton
Valves	drop
Crankshaft bearings	7 + 1 journal
Carburetor	1 Solex 40 JFF II
Firing order	1-5-3-6-2-4
Starter	Bosch BNG 2.5/12 + AL/ZMD
Generator	Bosch GTLN 600/12-1500
Batteries	1 12-volt, 120 Ah
Fuel supply	Pallas pump
Cooling	water
Clutch	two-plate dry F & S PF 220 K
Gearbox	Maybach Variorex "VG 102128 H"
Gears	7 forward, 3 reverse
Drive wheels	front
Axle ratio	1:5.9 (intermediate gear)
Top speed	55 km/h
Range	road 220, off 130 km
Steering	MAN clutch, mechanical
Turning circle	
Suspension	torsion bars
Lubrication	high-pressure
Brakes made by	MAN
Brake effect	mechanical
Brake type	outer shoes
Footbrake works on	steering brake
Handbrake works on	drive
Wheels	pressed disc wheels
Vehicle track	1920 mm
Track on ground	200 mm, 96 links per track
Track width	300 mm
Ground clearance	290 mm
Oberall length	5650 mm
Overall width	2300 mm
Overall height	2600 mm
Firing height	2200 mm
Ground pressure	0.87 kg/cm^2
Chassis weight	
Total weight	11,500 kg
Load limit	2000 kg
Seats	4
Fuel consumption	road 100, off 150 liters/100 km
Oil consumption	depends on action
Fuel capacity	200 liters, one tank
Armor: hull front	30 mm
hull other	14.5 mm
body	14.5 mm all around
Upgrade	24 degrees
Climbing	420 mm
Wading	850 mm
Spanning	1750 mm
Armament	1 7.62 cm Pak 36(r) L/54.8 (30 rounds) + 1 MG 34 (900 loose)
Uses	Makeshift tank destroyer
Similar	7.62 cm FK. 296 (r) on *Panzerjäger* II D and E

7.5 cm-Pak 40/2 auf Fgst.PzKgfwg. II (Sd.Kfz. 131) "Marder II"

Type: "LaS 100"
Made by Fahrzeug- und Motorenwerke GmbH, ex-Maschinenbau Linke-Hofmann (Famo), Breslau and aWarsaw works
Developed by ALKETT, Spandau

Years built: 1942-44
Information source: D 651/50 of 12/1/1942

Notes: 1217 units built

Motor	Maybach "HL 62 TRM"
Cylinders	6 in-line
Bore	105 mm
Stroke	120 mm
Displacement	6191 cc
Compression ratio	6.5:1
Engine speed	2600 rpm
Engine power	140 HP
Power to weight	12.7 HP/ton
Valves	drop
Crankshaft bearings	7 + 1 journal
Carburetor	1 Solex 40 JFF II
Firing order	1-5-3-6-2-4
Starter	Bosch BNG 2.5/12 + Bosch AL/ZMA
Generator	Bosch GTLN 600/12-1500
Batteries	1 12-volt, 120 Ah
Fuel supply	Pallas pump
Cooling	water
Clutch	two-plate dry F & SS K 230 K
Gearbox	ZF SSG 46 Aphon
Gears	6 forward, 1 reverse
Drive wheels	front
Axle ratio	
Top speed	road 40, off 20 km/h
Range	road 190, off 125 km
Steering	MAN/Wilson clutch
Turning circle	4.8 meters
Suspension	leaf springs, longitudinal, single-wheel

157

Lubrication	high-pressure	Chassis weight	6800 kg
Brakes made by	MAN	Total weight	10,800 kg
Brake effect	mechanical	Load limit	1500 kg
Brake type	outer band, self-adjusting	Seats	3
Footbrake works on	steering brake	Fuel consumption	road 90, off 135 liters/100 km
Handbrake works on	steering brake	Oil consumption	depends on action
Wheels	road wheels 550 x 100-55, jack rollers 20 x 105	Fuel capacity	170 liters
		Armor: hull front	35 mm
Vehicle track	1880 mm	hull other	14.5 mm
Track on ground	2400 mm	body	14.5 mm all around, front double
Track width	300 mm	Upgrade	30 degrees
Ground clearance	345 mm	Climbing	420 mm
Overall length	6360 mm	Wading	800 mm
Overall width	280 mm	Spanning	1700 mm
Overall height	200 mm	Armament	1 7.5 cm Pak 40/2 (37 rounds) + 1 MG 34 (600 loose)
Firing height	1940 mm		
Ground pressure	0.76 kg/cm^2	Uses	Makeshift tank destroyer
		Similar	Test vehicle with 5 cm Pak 38

le. FH 18/2 *auf Fahrgestell* PzKgfwg. II (Sf) (Sd.Kfz. 124) "Wespe"

Made in Germany
Made by Fahrzeug- und Motorenwerke GmbH, Breslau and Warsaw (ex- Vereinigte Maschinenwerke
Type: "LaS 100"
Developed by Alkett, Berlin

Years built: 1942-44
Information source: Handbuch WaA, Page G 365, etc.

Notes: 682 units built

Motor	Maybach "HL 62 TRM"	Brake effect	mechanical
Cylinders	6 in-line	Brake type	outer band, self-adjusting
Bore	105 mm	Footbrake works on	steering brake
Stroke	120 mm	Handbrake works on	steering brake
Displacement	6191 cc	Wheels	road wheels 550 x 100-55, jack rollers 20 x 105
Compression ratio	6.5:1		
Engine speed	2600 rpm	Vehicle track	1880 mm
Engine power	140 HP	Track on ground	2400 mm
Power to weight	12.7 HP/ton	Track width	300 mm
Valves	drop	Ground clearance	340 mm
Crankshaft bearings	8 journal	Overall length	4810 mm
Carburetor	1 Solex 40 JFF II	Overall width	2280 mm
Firing order	1-5-3-6-2-4	Overall height	2300 mm
Starter	Bosch BNG 2.5/12 + AL/ZMA	Firing height	1940 mm
Generator	Bosch GTLN 600/12-1500	Ground pressure	0.76 kg/cm^2
Batteries	1 12-volt, 120 Ah	Chassis weight	6800 kg
Fuel supply	Pallas pump no. 62601	Total weight	11,480 kg
Cooling	water	Load limit	
Clutch	two-plate dry F & S K 230 K	Seats	5
Gearbox	ZF SSG 46 Aphon	Fuel consumption	road 90, off 135 liters/100 km
Gears	6 forward, 1 reverse	Oil consumption	depends on action
Drive wheels	front	Fuel capacity	200 liters
Axle ratio		Armor: hull front	18 mm
Top speed	road 40, off 20 km/h	hull other	14.5 mm
Range	road 140, off 95 km	body	10 mm all around
Steering	MAN-Wilson clutch	Upgrade	30 degrees
Turning circle	4.8 meters	Climbing	420 mm
Suspension	leaf springs, longitudinal, single-wheel	Wading	800 mm
Lubrication	high-pressure	Spanning	1700 mm
Brakes made by	MAN/FAMO	Armament	1 10.5 cm le FH 18/2 L/28 (32 rounds)
		Uses	light armored howitzer

Gefechtsaufklärer VK 1602 "Leopard"

Type: "VK 1602"
Made in Germany
Made by Mühlenbau und Industrie AG (MIAG) Amme-Werk, Braunschweig

Year built: 1942

Information source: Handbook WaA, page D 27, July 1942
Notes: only drawings were made

Motor	Maybach "HL 157"
Cylinders	V-12
Bore	115 mm
Stroke	125 mm
Displacement	15,580 mm
Compression ratio	6.5:1
Engine speed	3500 rpm
Engine power	550 HP
Power to weight	21 HP/ton
Valves	drop
Crankshaft bearings	7 roller
Carburetor	2 Solex
Firing order	1-12-5-8-3-10-6-7-2-11-4-9
Starter	Bosch BNG
Generator	Bosch GTLN
Batteries	4 12-volt, 105 Ah
Fuel supply	pumps
Cooling	water
Clutch	hydraulic
Gearbox	Maybach pre-selector
Gears	8 forward, 1 reverse
Drive wheels	rear
Axle ratio	
Top speed	road 60, off 30 km/h
Range	road 300, off 150 km
Steering	multiradial
Turning circle	on the spot
Suspension	torsion bars, transverse
Lubrication	high-pressure and central
Brakes made by	Südd. Arguswerke
Brake effect	mechanical
Brake type	disc
Footbrake works on	drive wheels
Handbrake works on	drive wheels
Wheels	box type
Vehicle track	2430 mm
Track on ground	3475 mm
Track width	650 mm
Ground clearance	510 mm
Overall length	6450 mm
Overall height	3270 mm
Overall width	2800 mm
Ground pressure	0.49 kg/cm^2
Chassis weight	
Total weight	26,000 pk
Load limit	1500 kg
Seats	4
Fuel consumption	road 170, off 340 liters/100 km
Oil consumption	depends on action
Fuel capacity	520 liters
Armor: front	60 mm
other	20 mm
turret front	80 mm
other	50 mm
Upgrade	40 degrees
Climbing	750 mm
Wading	1750 mm
Spanning	2500 mm
Armament	1 5 cm KwK 39/1 + 1 MG 42
Uses	heavy reconnaissance tank; not built

Bibliography

Bülke, Willi A., *Deutschlandsa Rüstung im Zweiten Weltkrieg*
Duncan, N. W., *Panzerkampfwagen I and II*
Guderian, Heinz, *Erinnerungen eines Soldaten*
Heigl, Fritz, *Taschenbuch der Tanks*
Icks, Robert J., *Tanks and Armored Vehicles*
Magnuski, Janusz, *Wozy Bojowe*
Mellenthin, F. W. von, *Panzer Battles*
Meier-Welcker, Hans, *Seeckt*
Munzel, Oskar, *Die deutschen gepanzerten Truppen bis 1945*
Nehring, Walther, *Die Geschichte der deutschen Panzerwaffe 1916-1945*
Oswald, Werner, *Kraftfahrzeuge und Panzer der Reichswehr, Wehrmacht und Bundeswehr*, Motorbuch Verlag, Stuttgart
Scheibert, H. & C. Wagener, *Die Deutsche Panzertruppe 1939-1945*
Senger und Etterlin, F. M, von, *Die deutschen Panzer 1926-1945*
Spielberger, Walter J. & Iwe Feist, *Armor Series 1-10*
Spielberger, Walter J., *Der Panzerkampfwagen I und seine Abarten 1933-1941*
Spielberger, Walter J., *Der Panzerkampfwaagen II und seine Abarten 1934-1944*
Stoves, Rolf, *Die 1. Panzer-Division*
Tornau, G. & F. Kurowski, *Sturmartillerie*

Meanings of Abbreviations

A (2)	Infantry Dept. of War Ministry	Krad	Motorcycle
A (4)	Field Artillery Dept. of War Ministry	Kr. Zgm.	Motor tractor
A (5)	Foot Artillery Dept. of War Ministry	KS	Fuel injection
A 7 V	Transport Dept. of War Ministry	Kw	Motor vehicle, or war vehicle
AD (2)	General War Dept., Section 2 (Infantry)	l, le	Light
AD (4)	General War Dept., Section 4 (Field Artillery)	L/	Caliber length
AD (5)	General War Dept., Section 5 (Foot Artillery)	Le. FH	Light field howitzer
AHA/Ag K	General War Dept., Vehicle Group	le. FK	Light field cannon
AK	Artillery Design Bureau	l.F.H.	Light field howitzer
AKK	Army Vehicle Column	le. I. G.	Light infantry gun
AlkW	Army Truck	le. W. S.	Light army tractor
ALZ	Army Convoy	LHB	Linke-Hoffman-Busch
AOK	Army High Command	l. I. G.	Light infantry gun
APK	Army Testing Commission	Lkw	Truck
ARW	Eight-wheel Vehicle	LWS	Land-water tractor
A-Typen	All-wheel Drive (fast type)	m	Medium
BAK	Anti-balloon Cannon	MAN	Maschinenfabrik Augsburg-Nürnberg AG
Bekraft	Fuel Section, Field Vehicle Office	MG	Machine gun
BMW	Bayerische Motoren Werke (Bavarian Motor Works)	MP	Machine pistol
Chefkraft	Chief, Field Vehicle Office	MTW	Personnel transport vehicle
(DB)	Daimler-Benz	n =	Revolutions per minute (rpm)
DMG	Daimler Motoren Gesellschaft (Daimler Motor Co.)	n/A, n.A.	New type
Dtschr. Krprz.	German Crown Prince	NAG	Nationale Automobilgesellschaft
E-Fahrgestell	Uniform chassis	(o)	Stock civilian vehicle
E-Pkw	Uniform personnel car	Ob.d.H.	Army High Commander
E-Lkw	Uniform truck	O. H. L.	Highest Army Command
FA	Field Artillery	O. K. H.	Army High Command
FAMO	Fahrzeug-und Motorenbau GmbH	Pak	Antitank gun
FF-Kabel	Field phone cable	P. D.	Armored Division
FH	Field howitzer	Pf	Engineer vehicle
FK	Fiend cannon	Pkw	Personnel car
Flak	Anti-aircraft gun	Pz. F.	Armored ferry
F.T.	Radiotelegraph	Pz. Kgfwg.	Tank
Fu	Radio	Pz. Spwg.	Armored scout car
Fu Ger	Radio set	(R)	Tracks
Fr Spr Ger	Radiotelephone	R/R	Wheel/track drive
g	Secret	(RhB)	Rheinmetall-Borsig
Gen. St. d. H.	Army General Staff	RS	Tracked tractor
Gengas	Generator gas	RSG	Mountain tractor
G.I.d.MV.	General Inspection of Military Vehicles	RSO	Tracked Tractor East
g. Kdos	Secret command matter	RV	Targeting communications
gp	Armored	s	Heavy
g. RS	Secret government matter	schg.	Running on rails
gl	Off-road capable	schf.	Amphibian
GPK	Gun Testing Commission	Sd. Kfz.	Special motor vehicle
(H)	Rear engine	Sf., Sfl.	Self-propelled mount
Hanomag	Hannoversche Maschinenbau AG	S-Typen	Rear drive (fast types)
Hk	Halftrack	SmK	Pointed shot with core
H.Techn.V.Bl	Army technical order sheet	SSW-Zug	Siemens-Schuckert-Werke-Zug
HWA	Army Weapons Office	s. W. S.	Heavy military tractor
I.D.	Infantry division	Tak	Antitank gun
I.G.	Infantry gun	Takraft	Technical Dept, Inspection of Vehicles
In.	Inspection	TF	Carrier frequency (radio-technical)
In. 6	Inspection of motor vehicles	Tp	Tropical version
Ikraft	Inspection of field vehicles	*Vakraft*	Test dept, field vehicle office (WWI), Test dept. of vehicle inspection (*Reichswehr* & *Wehrmacht*)
Iluk	Inspection of air and ground vehicles	ve	Fully interference-free
K	Cannon	v/max	Maximum velocity'
KD	Krupp-Daimler	V^o	Muzzle velocity
KdF	Kraft durch Freude (Strength Through Joy organization)	VPK	Vehicle technical testing commission
K. d. K.	Commander of motorized troops	Vs. Kfz.	Test vehicle
K. Flak	Motorized anti-aircraft gun	ZF	Zahnradfabrik Friedrichshafen
Kfz	Motor vehicle	ZRW	Ten-wheel vehicle
KM	War Ministry	WaPrüf, WaPrw	Weapon testing office
KG	Motorized limber	*Wumba*	Weapon and ammunition procurement office
(Kg)	Krupp	wg	Able to go on water
Kogenluft	Commanding General of air forces